Fingerprints of God

Fingerprints of God

THE SEARCH FOR THE SCIENCE OF SPIRITUALITY

Barbara Bradley Hagerty

RIVERHEAD BOOKS

a member of Penguin Group (USA) Inc.

New York

2009

RIVERHEAD BOOKS
Published by the Penguin Group
Penguin Group (USA) Inc., 375 Hudson Street, New York, New York 10014, USA • Penguin Group (Canada), 90 Eglinton Avenue East, Suite 700, Toronto, Ontario M4P 2Y3, Canada (a division of Pearson Canada Inc.) • Penguin Books Ltd, 80 Strand, London WC2R 0RL, England • Penguin Ireland, 25 St Stephen's Green, Dublin 2, Ireland (a division of Penguin Books Ltd) • Penguin Group (Australia), 250 Camberwell Road, Camberwell, Victoria 3124, Australia (a division of Pearson Australia Group Pty Ltd) • Penguin Books India Pvt Ltd, 11 Community Centre, Panchsheel Park, New Delhi–110 017, India • Penguin Group (NZ), 67 Apollo Drive, Rosedale, North Shore 0632, New Zealand (a division of Pearson New Zealand Ltd) • Penguin Books (South Africa) (Pty) Ltd, 24 Sturdee Avenue, Rosebank, Johannesburg 2196, South Africa

Penguin Books Ltd, Registered Offices: 80 Strand, London WC2R 0RL, England

The author gratefully acknowledges permission to quote from William R. Miller and Janet C'de Baca, *Quantum Change: When Epiphanies and Insights Transform Ordinary Lives* (Guilford Press, 2001).

Library of Congress Cataloging-in-Publication Data

Hagerty, Barbara Bradley.
Fingerprints of God: the search for the science of spirituality / Barbara Bradley Hagerty.
p. cm.
Includes bibliographical references and index.
ISBN 978-1-59448-877-1
1. Religion and science. I. Title.
BL240.3.H34 2009 2009003921
215—dc22

Printed in the United States of America
1 3 5 7 9 10 8 6 4 2

Book design by Marysarah Quinn

To Mom,
the finest person I know

Contents

What if you slept? And what if, in your sleep, you dreamed? And what if, in your dream, you went to heaven and plucked a strange and beautiful flower? And what if, when you awoke, you had the flower in your hand? Ah, what then?

Samuel Taylor Coleridge

Fingerprints of God

CHAPTER 1

Crossing the Stream

I REMEMBER THE MOMENT I decided to leave Christian Science. It was a Sunday afternoon in February 1994. By my bleak accounting, New Haven, Connecticut, was enjoying its seventeenth snowstorm of the winter. I was completing a one-year fellowship at Yale Law School. I had abandoned my sunny apartment in Washington, D.C., for a dark cave in the Taft Hotel with rented furniture that, I'd realized upon delivery, was identical to that favored by Holiday Inns.

I was sick—sick with stomach flu, a fever and chills that induced me to pile every blanket, sweater, and coat in the apartment on top of me. Still I shook so violently that my teeth chattered. I slipped in and out of consciousness all afternoon, but in a moment of lucidity I envisioned the medicine cabinet above the bathroom sink. In normal circumstances, my medicine cabinet contained nothing more therapeutic than Band-Aids. I had been raised a Christian Scientist, and at the age of thirty-four—with the exception of receiving a vaccine before my family traveled to Europe—I had never visited the doctor, never taken a vitamin, never popped an aspirin, much less flu medicine. At that moment, what flashed in my mind's eye like a blinking neon sign was *Tylenol, Tylenol, Tylenol.* A friend of mine, I recalled, had left some Tylenol during a visit.

The bottle of Tylenol called seductively, and I followed its siren call. I slipped out of bed and, steadying myself on the furniture lest I faint, crept to the medicine cabinet. Before I could stop myself, I downed one tablet, closed the cabinet, averted my eyes from the mirror, and stumbled quickly back to bed. Five minutes passed. My teeth stopped chattering. Another minute or so, I began to feel quite warm, no, hot, *hot*, what was I doing under all these covers? I threw off the coats and sweaters and blankets and felt the fever physically recede like a wave at low tide. *Wow,* I thought, *I feel terrific!*

It was not thirty minutes later when I was up for the first time in two days and cheerfully making myself some tomato soup; it was not then, precisely, that I incorporated medicine into my life. It would take me another sixteen months before I would leave the religion of my childhood for good. Soon thereafter, I announced this to my friend Laura one day at lunch.

"Oh, Barb," she exclaimed, squeezing my hand in excitement. "Now the whole world of pharmacology is open to you!"

And so it was. But three decades of religious training does not evaporate quickly. As a Christian Scientist, I had come to believe in the power of prayer to alter my experience, whether that be my wracking cough or my employment status, my mood or my love life. In that time, I had witnessed several healings. I had come to suspect that there exists another type of spiritual reality just beyond the grasp of our human senses that occasionally, and often unexpectedly, pierces the veil of our physical world. In Christian Science we called these "spiritual laws," and (I was told over and over) I needed merely to bring myself in line with those higher laws to banish the cough or the heartache.

I say "merely," but it's actually tough sledding, trying to fix all your problems through prayer. In my mid-thirties, I chose the ease and reliability of Tylenol over the hard-won healings of Christian Science. More than that, I was tired of my ascetic diet of divine law and spiritual principles. I suppose I could have walked away from religion altogether, dismissing God and swatting away questions about eternity. But for whatever reason—my genetic wiring or the serotonin receptors in my brain or the stress hormones in my body—I held fast to the idea of God, of a Creator above and within this messy creation called my life and yours. I remained open to something unexplainable, even supernatural. But I did not have a clue as to how radically my life would be upended when I encountered that mystery one summer evening in Los Angeles.

ON JUNE 10, 1995, Kathy Younge and I were sitting on a bench outside Saddleback Valley Community Church. The Saturday-night service had ended an hour earlier. Even the stragglers had gone home. I was interviewing her for a *Los Angeles Times* Sunday magazine article about fast-growing churches—specifically, why baby boomers in their thirties and forties were flocking to evangelical churches. This took me into new spiritual territory. As a Christian Scientist, I had absorbed Mary Baker Eddy's version of Deity. The flinty founder of Christian Science defined "God" as a list of qualities—Life, Truth, Love, Spirit, Soul, Mind, and Principle. The Christian Science God is not a person. But to the evangelical Christians I met during my research for the *Times*, God is first and foremost a Person, one who came to earth two millennia ago and still yearns for a relationship with every human being.

Kathy Younge was my tour guide through this evangelical world. I was drawn to her because we occupied the same lonely demographic: both in our mid-thirties and single, we were wrestling with existential questions. But my questions paled next to hers. This woman had been fighting cancer for years. Her melanoma had recently returned, driving her to her knees in despair, and eventually to the comfort of Saddleback Church. Saddleback and its pastor, Rick Warren, are now almost household names, but in 1995, Rick Warren was unknown outside evangelical Christian circles, and his church drew only a few thousand people a week. (Now it's closer to 20,000.) Many of those people were like Kathy—broken in some way, physically, emotionally, spiritually, and famished for a living, breathing God who listens and intervenes. Saddleback is founded on this kind of God, and gives Him a structure to work with—twelve-step programs, ministries for every sort of physical, emotional, or financial challenge, and a massive prayer chain in which hundreds of Saddleback members pray for those in distress, like Kathy.

"Do you think the prayer group in the church will heal the cancer?" I asked Kathy that night, scribbling notes in the fading light.

"No. Healing comes from God," she said. "The church is here to be your family. They're really your support team down here because we don't have Jesus around to touch and talk to us. The church is God with skin on."

That was the quote that appeared in my *Times* article. What happened next did not.

"Kathy, how can you possibly be so cheerful when you've got this awful disease?" I asked.

"It's Jesus," she said. "Jesus gives me peace."

"A guy who lived two thousand years ago?" I asked, incredulous. "How can that be?"

"Jesus is as real to me as you are," she explained. "He's right here, right now."

Right, I thought. Yet there was something wondrous about Kathy's confidence as she struggled through this disease that could kill her. She told me then how she had been diagnosed with melanoma in her twenties, how her fear and loneliness had led her to Saddleback on a random Sunday, how she had come to believe that God had placed cancer in her life not to snuff it out but to give it a transcendent purpose.

As we talked, the night darkened. The streetlamp next to our bench cast a circle around us, creating the eerie sense that we were actors in a spotlight on a stage. The temperature had dropped into the fifties. I was shivering but pinned to the spot, riveted by Kathy and her serene faith.

My body responded before my mind, alerting me to some unseen change, a danger perhaps. I felt the hair on the back of my neck stand on end, and my heart start beating a little faster—as it is now, recalling the moment. Imperceptibly at first, the air around us thickened, and I wondered whether a clear, dense mist had rolled in from the ocean. The air grew warmer and heavier, as if someone had moved into the circle and was breathing on us. I glanced at Kathy. She had fallen silent in mid-sentence. Neither of us spoke. Gradually, and ever so gently, I was engulfed by a presence I could feel but not touch. I was paralyzed. I could manage only shallow breaths. After a minute, although it seemed longer, the presence melted away. We sat quietly, while I waited for the earth to steady itself. I was too spooked to speak, and yet I was exhilarated, as the first time I skied down an expert slope, terrified and oddly happy that I could not turn back. Those few moments, the time it takes to boil water for tea, reoriented my life. The episode left a mark on my psyche that I bear to this day.

EVER SINCE THAT NIGHT, I have wanted to write a book that answers the questions that I never voice in the two worlds I inhabit. The golden rule of journalism decrees that you take nothing on faith, that you back up every line of every story you write with hard evidence. You question everything. The unspoken ethos of organized religion is that you leave the uncomfortable questions alone, you accept them as unsolved mysteries or previously answered by religious minds greater than yours. You rely on the wisdom of sacred texts and your minister, and you swallow your doubts.

And yet I could not keep the questions at bay. Is there another reality that occasionally breaks into our world and bends the laws of nature? Is there a being or intelligence who weaves together the living universe, and if so, does He, She, or It fit the description I have been given? I was not worried about losing the old man with a beard—but what about the young man on a cross? Is there a spiritual world every bit as real as the phone ringing in the kitchen or my dog sitting on my foot, a dimension that eludes physical sight and hearing and touch? In the end, my questions boiled down to five words: *Is there more than this?*

As the years unfolded, I stockpiled more questions and more stories—strange stories that demanded an explanation. I wondered about my friend John, who was a slave to painkillers and scotch. Day in and day out, the cravings drove him away from his wife and to bars and Internet pharmacies. One day he felt the touch of something supernatural, and the cravings vanished. He stopped drinking and taking drugs, and while he never really bought into the tenets of his Catholic Church, he subscribed to the mysterious power that pulled him from the pit.

I wondered about my grandmother, a Christian Science "practitioner" or healer, who prayed for people and saw them recover. I thought about one patient in particular, a teenager in San Francisco who, tripping on LSD, jumped from an eighth-story window onto the street. The doctors declared him brain-dead and left him to Granny's prayers. Two weeks later, he walked out of the hospital.

I wondered about my own brief brushes with something numinous, and the gnawing suspicion that there might be a reality that hides itself except in rare moments, or to rare people. I have stumbled a few times into a mystical presence—once, as a blinding light in my bedroom, another time as a voice, several times as an undeniably physical presence. Few people know about those times; they are intensely private and, to be honest, a little bizarre. Which is why, I suppose, it took me more than a decade to write this book.

. . .

MARCH 6, 2004, exploded with promises of spring. I had spent the past two days moderating a forum on the history of faith and law in a hotel conference room in Walland, Tennessee. It had been heavy intellectual lifting, and when we finished the last session late in the afternoon, I burst outside like a kid at recess, sprinting to my hotel room and throwing on shorts and a sweatshirt. The hotel sat at the base of the Smoky Mountains, and my forty-four-year-old heart raced as I made a dash for the trail that led me into the nearest set of hills.

The sun was already hovering just over the horizon. But I only planned to hike a three-mile loop. Plenty of time. As I left the road and started up the trail, a longing swept over me, a memory of what it felt like to be young. Perhaps it was the smell of the spring moss, or the way the shadows dappled the path, but my mind's eye scanned back twenty-five years to those summers I had spent backpacking in the Colorado Rockies, in that pink dawn of life when your future fans out before you and you instinctively know that *now* is the time to risk everything because you may never have another chance.

The exhilaration did not serve me well. It blinded me to my surroundings and the stealthily setting sun. Only after I had hiked up and down one of the foothills did I notice it was growing quite dark. In the mountains, night dims gradually for a while—and then instantly, like a towel dropped over a birdcage, it is black. I started to run along the path that I knew would lead me back to the main road, when the trail simply ended. Curving from the left roared a small churning river, swollen from melted snow. On the right, the hill swept up vertically, thick with trees and prickly bushes. There was no path forward.

Perhaps, in retrospect, I should have waded into the ice-cold river and followed it to the main road. But doubts crowded in. *Maybe I took a wrong turn. Maybe I misread the map and this stream doesn't return to the road.* I decided that the only way back to safety was to retrace my steps.

For the next three hours I raced up and down the densely forested hills, blindly stumbling down one trail after another, unable to see my hand in front of my face. At some point, I became genuinely frightened. The Smoky Mountains are home to bears. A few days earlier, I had been warned (wrongly, I later learned), a serial killer had been caught in these very woods. Finally, I sat down, closed my eyes, and prayed. I prayed to a God who engages with His creation, a personal God intensely interested in every one of us, an all-seeing God whose GPS would guide me off the mountain. I began to sing Psalm 139, sotto voce at first, then louder in case the All-Hearing didn't pick it up. *"Whither shall I go from Thy Spirit, and whither shall I flee from Thy presence . . ."*

This is the image I hold in my mind: singing a psalm at the top of my lungs, clapping my hands to scare away the bears between verses. Bolstered by the prayer, I ventured down one more trail, and in a few moments I saw—*Thank God!*—lights from houses in the far distance, with people inside, tucked in for the night with a book. I careered down the rocky trail, relieved that my small nightmare was over. It was a perfect story for telling at dinner: a *little* dangerous but not too.

But the scene I was already imagining, regaling my dinner partners with this small adventure, abruptly ended. Between me and the little white house that had become my passage out of these black woods rushed a torrential river. Admittedly, it was more like a stream. But in the dark, it was a loud and angry stream, and dangerous, too, swollen to about forty feet wide by melted snow. Carefully, I waded into the water, hanging on to a tree branch that extended over the stream. The current instantly swept me off my feet. I clung fast to my sturdy little branch, surprised and grateful that it did not snap. Scrambling back onto the bank, I imagined some poor fisherman finding me three days later, miles downstream, caught in the brambles, my bloated body slapping softly against the bank.

Gazing at the racing water, I listened as carefully as I could. It sounds childish now, but at that moment the existence of God hung in the balance. At my core I believed I would hear a "voice" telling me what I should do. For several minutes, I strained to hear it. Nothing. Nothing but my own sullen admission that I had but one choice. I could stay on the mountain or cross that river.

And so I stumbled into the current as far as I could stand, and then, with a huge inhalation and a small cry, I heaved myself into the stream, tumbling and flailing until, miraculously, I smacked into the bank on the other side. I staggered out of the river, up to the little white house.

"Hello! Hello!" I yelled, banging on the door as I watched the lights turn out in the back. Then, timidly, a frail, white-haired woman peered out her window.

"Please," I begged, "I've been lost in the mountains for *hours*!"

"Oh, you're all wet," she said softly, opening the door a sliver. "Come in, come in."

I began to sob—relieved, yes, safe, yes—but feeling *alive* in a way I had not felt in years.

TRANSFORMING INSIGHTS USUALLY COME in small moments and pedestrian crises. So it was for me. In those few hours alone in the Smoky Mountains, in the dark soil of my fear, a seed split open. I realized that for nearly a decade, I had been running down one trail after another, keeping to the familiar safety of a path that would leave me dry but never lead me home. At some point, I needed to cross the river and immerse myself in the unnerving questions about God, and reality, and whether what I instinctively believed was true—or rubbish.

I was, to be honest, skittish. Skittish about ruining my reputation in a career where few people believe in God and fewer still bother to distinguish spirituality from religious politics. More than that, I was skittish about submitting my faith to scientific tests, exposing it to the possibility that the most profound moments of my life were nothing more than, say, electrical activity in my brain.

But in the end, I had to cross the river, and on the other side I found a small, brave band of scientists who had far more to lose than I did. They had slaved away for their medical degrees or their Ph.D.'s in neurology or biology, building a reputation in mainstream scientific research. And then they had risked it all by pursing the taboo questions. Is spiritual experience real or delusion? Are there realities that we can experience but not necessarily measure? Does your consciousness depend entirely on your brain, or does it extend beyond? Can thoughts and prayers affect the body? And that question I cannot seem to escape: *Is there more than this?*

Every generation claims a few scientists at the fringes wrestling with these questions. Sometimes they are called parapsychologists, a demeaning title that makes them sound illegitimate if not a little bit unhinged. But today's iconoclasts have an advantage their predecessors lacked. They have technology. They can peer inside a brain as it meditates in prayer or trips on psilocybin. They can look for markers in the brain, and, like forensic detectives, they are studying the evidence left behind by "spiritual" events that occurred out of their eyesight. They are trying to discern the fingerprints of the one—or the One—who passed through a person's psyche and rearranged his life. They are analyzing these "spiritual" moments, in the form of epileptic seizures or psychedelic experiences, meditations in a brain scanner or out-of-body experiences. In the process, they find themselves in a world of mystery.

I have been privileged to spend time with some of these scientists, as well as the people whose stories they are studying. Along the way, my own spiritual journey has taken a surprising turn. I have shed some beliefs as untenable. And I have reclaimed some beliefs I had long ago discarded—because science may be proving them true, or at least plausible.

I wrestled with the approach to this book. As a journalist, I naturally gravitate toward the safe, clinical, third-person approach to the science of spirituality, setting aside my personal predilections and laying out the evidence for the reader to evaluate. And yet, the questions in this book about the nature of God and reality—these are *my* questions, this is a *personal* quest, and therefore I have woven my own story into the larger quest.

Like anyone else who has lived nearly five decades, I am hardly a blank canvas. My life experiences have nudged me into my particular perspective. But I have tried to draw on two experiences to help steer this book. The first is my quarter-century as a print, television, and radio journalist. I habitually shelve my preconceived ideas to try to glean the truth. I have learned to ask questions and really listen to the answers. If the science surprises me, or if it simply does not fit into my worldview, I must take that into account if I am to be an honest guide. Even if the science contradicts what I feel in my gut to be true, I will relay it to the best of my ability.

The second influence is my particular faith journey. Even though I now consider myself a mainstream Christian, I have never entirely shed the perspective of Christian Science. Christian Scientists do not evangelize or impose their truth on anyone else, and they are graciously tolerant of all other faiths. I am not going to foist my findings on you, and I will certainly not advocate a particular religious view. I wrote this book because some questions plagued me. I set about answering them—not by gazing at my navel, for there is only so much information that any one navel can provide. Rather, I am examining the science and the scientists, the "neurotheology," that might explain those numinous moments that most everyone has enjoyed.

THIS BOOK TACKLES the existence of God, "the reality of the unseen," as psychologist William James had it. After talking to countless scientists far more knowledgeable and insightful than I, I have concluded that science cannot prove God—but science is entirely consistent with God. It all depends on how you define "God." If you are trying to locate deity in a thirty-three-year-old carpenter or the unseen divider of the Red Sea, science will offer no help. But if you look for God in the math of the universe, if you perceive God as the Mind that rigged existence to create life, then science can indeed accommodate. If you see God in the breathtaking complexity of our brains, as the architect of our bodies and our minds who planted the question *Is there more?*—well, science has room for that kind of God.

This is the God I am pursuing in this book. If I were to draw a road map for this journey, I would say the first part is driven by personal questions about my own spiritual experiences. Chapter 2 explores spontaneous spiritual experiences that come unbidden, like the one that washed over me as I sat talking with Kathy Younge. It turns out I was not alone: half of Americans have been overcome and radically transformed by a sudden encounter with the spiritual. Science is only now catching up with the ideas of William James, ideas that it has discarded for a century but can no longer ignore.

In chapter 3, I raise the question at the foundation of my own religious upbringing: Is there a God who hears prayer and heals? When you look at massive prayer studies involving hundreds of people, the evidence for prayer's efficacy is mixed at best. But go beyond the blunt instrument of strangers praying for strangers, look for the power of thought to stop the HIV virus in its tracks, for example, and suddenly the picture changes. What was once superstition is now accepted science: our thoughts affect us on a cellular level, unfolding the biology of belief.

Chapter 4 looks at the triggers for spiritual experiences. Is there a certain set of circumstances, a certain personality type, a certain cocktail of internal and external stress, that erupts in a spiritual experience? I believe there is, and I believe this explains why alcoholics so often become spiritual people. While an encounter with God can happen anywhere, anytime, my research and my own life story tell me that *brokenness* is the best predictor of spiritual experience.

Next, I investigate the hard, arid, materialist science that tries to explain (away) spirituality. In the process, I stumble on a new definition of *deity*.

In chapter 5, I ask the question *Why me?* Why are some people inclined to pursue and experience God while others couldn't care less? Is there a "God gene" that predisposes a person to Spirit? It seems to me that one way to define God is as a master craftsman who organizes our genetic code so human beings have a capacity and yearning to know Him.

But God has other talents, and other roles. Chapter 6 explores God as chemist, who adjusts the chemicals in our brains so that we can gain access to the spiritual. To find this God, I traveled to a peyote ceremony in Arizona, and to Johns Hopkins University, where a prominent neuroscientist has found that psychedelic drugs hold a key to understanding our connection with the spiritual.

Chapter 7 reveals God as electrician, who wires our brain to allow us to tune in to an unseen reality. For this I visited an epilepsy clinic in Detroit. Scientists have long believed that the ancient mystics like Saint Paul and Saint Teresa of Ávila did not experience God but only the electrical storms of temporal lobe epilepsy. Recently, however, some neurologists have begun to speculate that those neurological events may not spark mere delusions—but actually allow people to hear and see a spiritual dimension that normal consciousness cannot grasp.

In spontaneous mystical experiences, people stumble into the ineffable; but knowing God is also a muscle that one can develop. Chapter 8 follows spiritual virtuosos—people who practice accessing God as intently as Tiger Woods perfects his golf swing. I witnessed a Christian virtuoso as he meditated in a brain scanner at the University of Pennsylvania, and tried my own hand at altering my brain through spiritual practice. Along the way, I began to question two exclusive theologies—the Christian doctrine that there is only one way to God, and the materialist doctrine that there is no God at all.

Finally, I find myself in territory just beyond the boundaries of "acceptable" science, and here I discover some tantalizing explanations for how science and spirit might coexist. In chapter 9, I look at the twentieth-century assumption that the brain is the sum total of one's consciousness. Once the brain stops functioning, most scientists say, so does one's mind, along with one's identity and existence. Out-of-body experiences fly in the face of this assumption, including the remarkable—and clinically documented—case of a woman who had no brain function but continued to think and observe.

Chapter 10 tackles a different but related question: What happens to people who touch death and return? Today neurologists are submitting people who have had near-death experiences to scientific scrutiny, hooking them up to EEGs and sliding them into brain scanners. The results of these experiments do not *prove* that we have souls that survive death—that is beyond the capacity of science to determine. But there is room in this science to believe in another reality beyond the veil of death.

In chapter 11, I come to believe that a major problem with God is not whether He exists but how one defines Him. Indeed, scientists from Albert Einstein to Stephen Hawking have looked at the universe and acknowledged a being, or mind, or architect before whom the only response is awe. The "God" that scientists can accept, in His most recent iteration, springs in part from quantum physics, the mysterious behavior of the very smallest particles. I caught a glimpse of this physics in action at a research institute in California, where I watched as one person's thoughts appeared to affect another person's body. This, then, is where mysticism meets science, and points to a God who may be the medium through which our prayers and thoughts are transmitted to others.

In chapter 12, I find two sorts of paradigm shifts. One, I believe, is just beginning to take place in science. The other has already begun to take place in my own soul.

Now let's get on with this radical project. But before we do, I want to make one more observation. Belief in God has not gone away, no matter how secular society has become or how much effort reductionist science has exerted to banish Him. God has not gone away because people keep encountering Him, in unexplainable, intensely spiritual moments. That is where I begin this scientific hunt, with the people who intimately know this God.

I begin with the mystic next door.

CHAPTER 2

The God Who Breaks and Enters

If I had seen Sophy Burnham in the produce section of my local grocery store I would not have thought, *Oh, she looks like a mystic.* But that is precisely what she is—a woman who has perforated the veil of physical reality and glimpsed another world. She lived only a few blocks from me, and after reading her book *The Ecstatic Journey*,[1] I made an appointment to see her. I approached this interview with a sort of fascinated terror. I was drawn to the untamed, almost uncontrolled spirituality she had described in her book. But I was also skittish because her journey bore an unsettling resemblance to my own.

True, her mystical experience was a nuclear explosion compared with my little blast of dynamite. But I knew instinctively that I had the same sensibility, the same sort of internal compass that might set me on a parallel spiritual path. This haunted me. I recoiled at Sophy Burnham's adventurous faith life, the way she bushwhacked through the spiritual woods, testing this religion and that. I was quite comfortable with my faith life, with the rhythms of a certain kind of prayer and reading each morning, a certain type of church and pastor, a reliable faith where I could turn a knob to get hot water from the faucet or a flame from the stove. For all the excitement, Sophy had paid an exorbitant price for her spiritual journey. She had ended her marriage, given up her longtime friends, and abandoned her comfortable life as a Washington journalist. I glimpsed in Sophy's past my own potential future, and it scared the hell out of me.

The woman who ushered me into her Washington apartment was small and lithe, wearing black pants and an orange crocheted sweater. She bustled off to the kitchen, her movements quick and precise as she prepared a cup of herbal tea. Sophy Burnham was nearing seventy, but looked twenty years younger. She wore her dark-blond hair short, simple, and parted on the side, framing a delicate nose and mouth and—clichéd as it sounds—penetrating eyes. She gave me a brief tour of her airy apartment, which occupied one of the city's oldest and most elegant buildings. The tall windows overlooked a canopy of lush trees on this morning in May. The eclectic artwork on her walls reflected the taste of a woman who has traveled widely: two large oil paintings of saints (booty from the Mexican War); a framed page from a medieval manuscript; an Indian cloth painting of a blue Lord Krishna with his gopis, cows, and peacocks. She cleared a space on a dining room table cluttered with manuscripts and began her story.

Sophy was three or four years old when she first felt some kind of "connection with the universe."

"It wasn't that there was something more than this—that there's a physical world and something else," she explained. "I had the sense that—like Gerard Manley Hopkins—the physical world was glowing with life. Shining like shook foil. I knew that everything was alive, and that the universe was on my side."

When she was ten or eleven, she was out riding with her father on their remote Maryland farm.

"My father said to me, 'For some reason the horses are really spooking.' Because they were dancing all over the place. And I said, 'Well, of course they are, they're feeling all this electricity in the air.' He said, 'Oh? Is there electricity in the air?' And that was the first time that I realized, Oh, not everybody feels this. Isn't that interesting? I just assumed that everybody was intimately attuned."

I wrote down her comment and circled it in my notes. As I talked with more people, I would find that mystical adults were once mystical children, as if they were genetically wired for the spiritual.

Over the next thirty years, Sophy would attend Smith College, stray from her Episcopal upbringing, and dabble in atheism. She would marry a *New York Times* reporter and give birth to two girls. In her thirties, the part of her psyche that was a spiritual nomad broke free. She began to seek answers from a Hindu guru and Buddhist meditation. Sophy arrived at age forty-two with teenaged girls, a caring husband, a glittering social life, and a demanding career as a freelance magazine writer.

"I was happy, and yet there was something deeply missing," Sophy said. "And it was a deep, deep longing that couldn't be satisfied. I remember looking in the mirror and thinking, *Is this all?* And then thinking, I have everything. I have a loving husband and a house and children and friends and a career. Why am I yearning for something else? And I didn't know what I was yearning for."

She found it on assignment. *Town & Country* magazine sent her to Costa Rica and Peru to profile the World Wildlife Fund. As an afterthought, she added a side trip to Machu Picchu, the sacred Inca mountain ruins in Peru.

She was sitting in the airport in San José, Costa Rica, when she had a premonition of the mystical experience to come.

"Suddenly, I saw everything shining, shining, and the *people* were shining," she recalled. "Everyone was luminous, and it was so profound that I just sat at my table, I just shook, with tears streaming down my face, it was so beautiful. And I thought, *Oh well, that's it. That's as good as it gets.*"

She was wrong. Two days later, when she was climbing the terraces of Machu Picchu with a group of other tourists, "I felt a nagging little chord in me, saying, 'You've got to go away, you've got to go away. You don't have much time. Hurry. *Hurry.*' "

She left her companions and scrambled up to the terraces, where she could be alone. She sat down, closed her eyes, and instantly was in "another place."

"The first thing that happened was frightening," Sophy said, looking into the middle distance. She spoke slowly, carefully, as if she were narrating an event she was witnessing in that very moment. I can do justice to her story only by repeating it verbatim.

"The first thing that happened was the sound of a hollow darkness. I've never heard it except there. It was very, very big—sort of like an oncoming train—but I knew it was in my ears, it was not external. And then there was a hand at the back of my neck, pressing me very strongly down. Everything is dark. Everything is black. And a voice that said, 'You belong to me.' And my response was, '*If* you are God. I belong to God.' And immediately everything turned to light."

She paused for ten, fifteen seconds, as if absorbing the light.

"The rest of the time, I was captivated in this mystical revelation. I was shown things that I don't even have the wit to ask questions about," Sophy said. I thought of Saint Paul's vision of third heaven, when he claimed he was "caught up in paradise," and "heard inexpressible things, things that man is not permitted to tell."[2]

"I don't remember most of it now," Sophy said, reeling me back to the moment. "But I do remember two things very vividly. One was the sense of seeing the beginnings and ends of the world—how it all began. And the other was this image—it's very difficult to describe—of electrons or atoms being swept along in the path of the hem of the garment of God. I think it was the passing of the Holy Spirit, which I could not see, because to see that would blind me."

Sophy laughed self-consciously, embarrassed to place herself in the company of Moses watching the back of God from the cleft of the rock.

"And seeing the aftereffects! It was all love and joy and sparkling particles, swarming up and circling—it was just exquisite. And knowing that everything is going to be all right. It's that idea of Saint Julian of Norwich, although I didn't know it at the time: *'All shall be well and all shall be well and all manner of things will be well.'*[3]

"And then, slowly, slowly, slowly, coming like a turtle up to the surface of the water again, I opened my eyes. And I was blinded by the light. It was piercingly painful so I shut my eyes again. And I went back to the spiritual light for a while, and then came out again, and when I opened my eyes this time, I could see. And I realized the whole thing had taken forty-five minutes, a *huge* amount of time. And I realized I had to hurry for the bus. And I came galloping and springing like a gazelle down the terraces, my heart filled with joy! And watching this light radiate off my hands and light off my arms and light off the grasses and the trees burning with light, everything flaring!

"I suppose that's what God sees when he sees us," Sophy reflected, turning to look into my eyes. "Just light. Nothing else. And then I got on the bus and I noticed that a blood vessel had burst in the back of my hand about the size of a quarter. It startled me, but also pleased me because it meant that yes, something had happened. And other people knew it. I remember this university professor coming up to me and sitting next to me and saying something like, 'Something happened to you, didn't it?' I said, 'Yes,' and that was all."

I sat immobilized by her story. I felt like a body surfer who had been slammed down by a wave. Then I asked Sophy a question that many a neurologist has pondered as well.

"Did you think, *Gosh, I just had a temporal lobe seizure*?"

"Oh yes! Absolutely that occurred to me!" Sophy admitted happily. "Was that an epileptic fit? Did I have some kind of electrical burnout of the brain? But everything seemed to be functioning," she said, adding that nothing like this had recurred in the past twenty years.

She leaned forward, speaking urgently.

"The *experience* is not important," Sophy said, and then she laughed. "I've just spent fifteen minutes telling you about an experience and now I tell you—and I cannot reiterate it enough—*the experience is not what was important.* It's changing you on a cellular level that is important. It's providing the hope and joy that's important."

If a spiritual experience is real, she said, it will transform you, fling your worldview and priorities, your relationships and your personality, up in the air like a two-year-old hurling a deck of cards. It can make you a stranger to your friends, to your family, and to your own psyche. It was a disorientation I knew all too well. I remembered my own return to Washington after researching the *Los Angeles Times* story, when I, too, felt my world upended by a brief brush with something mystical. I remembered how my friends and even my family felt foreign, how I declined invitations to dinner so that I could stroll for hours at night, reveling in the unseen company of a God I had just discovered.

"What was it like coming home from Peru?" I asked.

"It was dreadful," she said quietly.

"But your life in Washington was so rich," I protested.

"Oh yeah, I'm a successful writer, I'm married to a successful journalist for *The New York Times*," she conceded. "But that was ashes in my mouth. I could not bear it. It was physically painful to sit at a dinner party and listen to the shallowness of the conversation. I was so sensitized. I could hear what was going on underneath people's conversations. This woman is telling a story at a dinner party in a brittle, gay, happy way, and underneath it, I can hear her heart breaking! I just wanted to shake people and say, *Stop it. I can't bear it!*"

I persisted, prompted less by journalistic curiosity than by a need for guidance, affirmation perhaps. "A lot of people would say, 'Yeah, that was a remarkable experience, but now I'm back in my real life in Washington, D.C., with my friends and we talk politics, we talk economics, we talk journalism.' Don't you think it's easier to fit back in that way?"

"That's true. Except that you're forgetting one thing," she said. "You've fallen in love."

This is the paradox of the person who is tethered to earth but has touched the sky. The memory robs *and* enriches you, reveals your life to be drabber *and* more magical than you had imagined.

"Anyone knows, when you're passionately in love everything is heightened," Sophy explained. "And you can find yourself in this wonderful state, the state of being in love that so many saints talk about when they speak of being the 'bride of God.' We always think it's metaphorical, but what it really means is that they're in a perpetual state of being in love. And just the way you can't scientifically measure being in love, although you know it's a madness, and that surely it will end—it's the same with a spiritual experience. It's a madness and it will die down and be replaced by something else."

Eventually, perhaps, but not without exacting its price. It is an imperious love. It usurps all rivals. It did for Sophy.

"I can remember my husband saying at some point, 'How can I compete with God?' " she recalled. "And I said, 'You can't.' "

Sophy Burnham tried to make her marriage work. She had teen-aged children, after all. But she had been transformed, as she put it, at a "cellular level"—to the core of her being, where she could think of little else but to pursue this God she had encountered on Machu Picchu. There was no turning back. It was as if the door back to a normal life had been locked from the other side, and she could only move forward. And like other modern-day mystics I interviewed, Sophy set about transforming her external life to match her new inner world.

The Burnhams divorced three years after Sophy's trip to Machu Picchu. Her daughters were furious for several years, but Sophy says she has repaired those relationships. Her choices—her single-minded pursuit of God and her resulting singleness—is yet another reason her story terrifies me.

Had Sophy Burnham been lying in a brain-imaging machine at the time of her mystical experience, a neurologist might explain the incident like this: The part of Sophy's brain that orients her in time and space became quiescent. Her spatial boundaries fell, creating the sense of oneness with the universe. Meanwhile, the part of the brain that handles hearing, vision, and emotions spiked, creating the roaring sounds and the particles of light that composed, for her, "the hem of the garment of God."

But just beneath that physical explanation lie explosive questions that scientists often prefer to ignore or to declare irrelevant. Who or what *causes* these spiritual dramas, these tiny gold threads of mystery that have woven their way down the centuries and into religions, through Christianity and Buddhism, through Islam and Kabbalah and Hinduism? Often, scientists can spot patterns in these mystical accounts and, with a sigh of relief, offer a diagnosis. Oh, that's temporal lobe epilepsy. It's schizophrenia. It's LSD, or magic mushrooms, or a chemical that is released in the brain as it shuts down to die.

It is, both literally and figuratively, all in the head.

And yet, small cracks are appearing in the smooth façade of this paradigm, thanks to a small army of psychologists, geneticists, and neurologists. They are making surprising discoveries about the physiological underpinnings of the spiritual. Before delving into DNA and brain chemistry, I want to turn to the most mysterious genre of experience—the spiritual storm that passes through otherwise healthy people, often unbidden, usually unexpected, and always unexplainable by material science.

This is the God who breaks and enters.

William James's Outrageous Ideas

If Sophy Burnham had crossed paths with William James, her story might have found its way into the pages of his *Varieties of Religious Experience.* Sophy missed the famous Harvard psychologist by a century:

James's series of lectures was published in book form in 1902. That *Varieties* is still regarded as the classic attempt to understand spiritual experience from a scientific perspective is, no doubt, a tribute to the originality of James's thinking. It is also a reflection, and perhaps an indictment, of twentieth-century science, which shied away from investigating this most basic of human sentiments—the longing for "something more."

I like to imagine William James arriving at the University of Edinburgh in 1901 to deliver the prestigious Gifford Lectures on Natural Religion. I can picture him, with full beard, receding hairline, and thick eyebrows crowning his intense eyes, approaching the lectern, gazing out at the sea of European scientists, philosophers, and intellectuals, and taking a deep breath.

"It is with no small amount of trepidation that I take my place behind this desk," he began his first lecture, "and face this learned audience."[4]

James professed to be intimidated by his colleagues' erudition. But he must have felt a twinge of exhilaration as he began to dissect, like a sure-handed surgeon, the philosophical wisdom of his age. In his twenty lectures, James rejected the reigning theory of his colleagues, who diagnosed spiritual experience as evidence of a disordered brain and believed that mystics and religious believers were better suited for the asylum than pulpit or pew. Why, he asked, could scientists not envision the world as consisting of "many interpenetrating spheres of reality,"[5] which can have both scientific and spiritual explanations—just as, today, depression can be explained by both psychotherapy and altered brain chemistry?

"The first thing to bear in mind," he warned the august crowd, "is that nothing can be more stupid than to bar out phenomena from our notice, merely because we are incapable of taking part in anything like them ourselves."[6]

Then James set out to do what scientists do so well. He categorized. First he analyzed the more common (and less intense) spiritual experience: religious conversion. Drawing from the rich personal stories of Saint Augustine, Leo Tolstoy, John Bunyan, and less famous converts, he concluded that people who had undergone religious conversions had achieved something that most of his intellectual colleagues had not. They had pushed through the angst of a "divided self," with its disappointments and doubts about the world. They had utterly surrendered themselves, only to find that their divided self had somehow been sewn back together.

For psychologists, James said, those forces are subconscious; but for the converted, they are supernatural. James himself yearned for, but never found, the transforming spirituality about which he spoke. But, he told his colleagues, he had become persuaded that "twice born" people—those who had been touched and converted by spiritual experience—were the healthiest, not the sickest, among us.

James then turned his gaze to the spiritual virtuosos: the mystics, who know firsthand "the reality of the unseen." All mystical experiences share certain common elements, he contended—and his descriptions presaged the words of Sophy Burnham. First, they are ineffable: they cannot be described adequately with words. Second, they have a noetic quality—a truth or deep insight that is truer to the person than the material world itself. The insights may differ from person to person, but there are certain shared threads: the unity of all things, the love of a conscious "other" or God, and the confidence that all is as it should be—as Julian of Norwich voiced it, "all manner of things will be well." Third, James observed, mystical experiences quickly ebb, usually within a few minutes. Last, these experiences pounce on the mystic: some power outside of the person takes control, pushing the mystic into the passenger's seat. As James put it, "The mystic feels as if his own will were in abeyance, and indeed sometimes as if he were grasped and held by a superior power."[7]

After placing mystics under his searching and sometimes critical gaze—for example, he looked dimly at Saint Teresa's erotic flirtation with God as so much exhibitionism—the Harvard scientist came to an extraordinary conclusion.

"Our normal waking consciousness, rational consciousness as we call it, is but one special type of consciousness, whilst all about it, parted from it by the filmiest of screens, there lie potential forms of consciousness entirely different," he said.

He went on to write, "We may go through life without suspecting their existence; but apply the requisite stimulus, and at a touch they are there in all their completeness, definite types of mentality which probably somewhere have their field of application and adaptation. No account of the universe in its totality can be final which leaves these other forms of consciousness quite disregarded."[8]

For many believers then and now, William James had endorsed "God." At the very least, he had vouched for the mental health of those who claim to know God. James might have viewed it differently: he readily conceded that science could neither prove the reality of a spiritual dimension, nor rule it out. But these experiences, he argued, point to "the possibility of other orders of truth."[9]

James's lectures caused a stir in academic circles around the world, drawing accolades and more than a few attacks. But if James triggered a revolution in the scientific approach to spiritual experience, that revolution was soon overcome by alternative theories that excluded the mystical.

One theory, launched by psychologist James Henry Leuba around the time of James, declared God to be illusion. In this view, electrical or chemical activity in the brain is the source of all mystical experience;[10] epilepsy and psychedelic drugs only prove the point. Sigmund Freud dismissed spiritual experience as pathology. Freud stated that religion offers protection against suffering through a "delusional remolding of reality." The "oceanic feeling" that mystics describe is a memory of an infantile state, he argued, perhaps unity with the mother—but whatever it was, it represented an escape from reality, not a perception of it.[11] Others went a step further, arguing that mystics longed for their mother or the womb, a desire sparked by the sexual frustrations of chastity. Piled on top of these psychological arguments was a sociological one: French sociologist Émile Durkheim[12] and others theorized that mystics are correct to feel part of something larger—but that something is society, not God.

True, a keen scientist in the twentieth century could, if he strained, hear faint echoes of those who believed that spiritual experience involved more than the byproduct of misfired synapses. Canadian psychiatrist Richard Bucke theorized that mystics had an evolutionary advantage that allowed them to tap into "cosmic consciousness," a spiritual octave too high for ordinary humans to hear.[13] Carl Jung believed that numinous or supernatural experiences are everywhere and invisible, like the air we breathe. He argued that each person can access a "collective unconscious." Mystics, then, are people who experience the collective unconscious vividly.[14] And in the 1960s, psychologist Abraham Maslow coined the term "peak experiences"—ecstatic moments of joy that occur when exceptional (or "self-actualized") people feel unified with the world and aware of ultimate truth. Maslow stated such experiences could be secular; but his work cracked open the door for the scientific investigation of mysticism.[15]

However, these were mere footnotes to the triumphant scientific idea of the twentieth century: namely, that science has no business sticking its nose into spirituality at all. Behaviorism, which was conceived by John Broadus Watson and popularized by B. F. Skinner, suggested that if a scientist cannot observe something directly, then it may not be real—and anyway, it is not relevant to science. From this perspective, if mental states and consciousness fell beyond the boundaries of scientific inquiry, then spiritual experience resided on another planet.[16]

Thus the scientific study of the spiritual was shoved to the side. The work of serious scientists was dismissed as little more than psychic voodoo. Careful experiments involving precognition and telepathy were thrown into the same ghetto as stories of haunted houses and UFOs.

And yet, the public refused to dismiss spirituality. Beginning in the 1950s, the power of positive thinking, and the growing evidence that thoughts (or prayer) can heal one's body, forced scientists to take a closer look at the mind-body connection. In the 1970s, sparked by Raymond Moody's book *Life After Life*, millions of people testified to experiencing life after death, suggesting we do have a soul. Some reputable universities, such as Princeton and the University of Virginia, began to study spiritual phenomena in earnest. While no one can pinpoint the moment when modern scientists began to take spiritual experience seriously, many say they know what sparked the new interest: technology. Neurologists in particular, armed with EEGs and brain scanners, were able to peer into the brain and witness spiritual experiences unfold. After a century, the tools of science caught up with William James.

Picking Up William James's Gauntlet

How I would have loved to talk with William James about the present-day research in "neurotheology"—the study of the brain as it relates to spiritual experience. But I found a proxy for James, one who had the advantage of being available for interviews. William Miller would turn out to be my guide through the psychological exploration of mystical experience, and like the other guides who led me through genetics, neurology, and quantum physics, Miller seemed delighted to spend a few hours talking about his research, challenging, as it does, the materialist assumptions of science.

"When I went to graduate school, I got the very clear message: Spirituality is *not* something it's okay to talk about," Miller recalled. "We could ask our clients about anything else. We could ask them about family life and violence in the home and sex life, but never about spirituality or religion. It was the great taboo."

Miller leaned forward in his chair. It was late July 2006, a blindingly hot day in Albuquerque. A lanky, bearded, rugged fifty-something, Miller looked more like a forest ranger than a Distinguished Professor of Psychology and Psychiatry at the University of New Mexico. The look seemed fitting for an academic with Miller's chosen path. His specialty was the psychology of spiritual experience, a path so long neglected that William James's footprints were barely visible for the underbrush. And yet, Miller has attained a sort of cult status among social scientists at universities like Harvard and the University of Pennsylvania. He was brave enough to tackle the supernatural mysteries of the psyche.

"William James was talking about these things in a very respectful way at the turn of the century," I said. "What happened between then and now?"

"My own metaphor is that psychology had to go through adolescence. Because psychology came from religion and philosophy, we had to say, 'I am *not* religion, I am *not* religion, I am *not* religion.' And then as you grow up, you kind of find your way back to your roots. And I think psychology is doing that: we're reconnecting with William James and the philosophic and theological roots that were there when psychology began."

My own view is that psychologists and other scientists have been ignoring the spiritual at their peril, since half of all Americans claim to have had some sort of spiritual experience. Psychology is the study of humans, after all, and at some point psychologists should probably accommodate these peculiarly human beliefs. More than 90 percent of Americans believe in God. Compare that with the scientists: recent polls suggest that only 40 percent of scientists overall and 7 percent of the elite scientists in the National Academy of Sciences believe in God.[17] Therein lies an explanation for why researchers have neglected the spiritual. Why should they study what they consider a delusion?

Bill Miller's own quest to study the spiritual began in the most personal of ways in the summer of 1989. Miller and his family were preparing for a yearlong sabbatical in Australia. All of the family but one. His fourteen-year-old adopted daughter, Lillian, had gone through a particularly rebellious period and was living at a girls' ranch two and a half hours away. One day, Miller trekked off to visit her, only to find that she refused to see him.

"So I turned around and drove home again," he said. "And then later that week she called, and on the phone, her voice was different. There was something different about this voice I was hearing. And she said, 'Dad, I need to talk to you.' "

He arrived at the ranch a few hours later.

"And this was a different person from the one I had known before that. And I said, 'What happened?' And she said, 'There's a mandatory chapel here. I went and sat in the back row like I always do, waiting for it to be over. And then this *thing* happened to me and I don't even know how to talk about it. I don't have any words to describe it.'

"She's really been a different person ever since," he continued, noting that his daughter was now thirty. "And I won't say she's been without problems or struggles or difficulties. But in the sense of being a different, compassionate, empathic person with a sense of reality bigger than the material world—yes, that came that day and has been there ever since."

If this had been a one-off, Miller might have let it slide away. But for some time, he had been hearing about these epiphanies from clients in his clinical work at the university. He noticed that some of his patients underwent sudden spiritual experiences, and when they emerged on the other side, they had been transformed: no longer alcoholic, no longer suicidal, they were people who treated life as a gift. He called the phenomenon "quantum change."

"And I couldn't think of any theory, or even a word, that described this kind of change within my own discipline," Miller said. "That was part of what fascinated me. I thought, *Wait a minute, if a personality can change literally overnight, and it's permanent, shouldn't I be interested in that as a psychologist?*"

Miller decided to pursue the question, and to find his subjects, he persuaded a journalist at the *Albuquerque Journal* to write a story about his work. At the end, the article noted that Miller was eager to talk with anyone who had undergone a dramatic psychological change or spiritual epiphany.

"The next day, the phone rang and rang and rang," he laughed, still tickled by the memory. "In the end, eighty-six people called us, of whom fifty-five agreed to come in and spend three hours with us for no pay, to just tell us what had happened to them. And it turned out one didn't need sophisticated clinical interviewing skills. You just had to say, 'Tell me about your experience,' and it would just come pouring out."

Those stories fill the pages of Miller's book *Quantum Change.*[18] Many times, the event was subtle yet caused a wholesale shift in the person's approach to life. Miller recalled the mother whose teenaged daughter had once again slipped out in the middle of the night. She was paralyzed with anxiety, but this time she surrendered her sickening fear to God. In the next instant, "there was just sort of a wave of knowing that everything was going to be all right," she told Miller. "All of a sudden the desperation was gone and in its place was this assurance. . . . I felt at peace. I felt reassured. I felt loved, too, like there was a union; I was a part of something that was a loving, peaceful thing, something a lot bigger than me."[19]

Other times, the mystical event crashed into the person's life, often in the middle of a traumatic event, as happened to the woman whose abortion went violently wrong. Instantly she found herself transported out of her body.

"I saw heaven and I saw hell," she told Miller. "I also knew that there is much more than the physical body and that when you die, you go on. . . . And then I just felt so at peace. It was very, very, very calming, because I knew something really important. I knew I was no longer limited by the physical. . . . I was able to transcend the physical, to see beyond the physical limitation. I have never feared death since then, nor have I ever experienced depression, which was a major change for me."[20]

My favorite story involved a scientist whose alcoholism landed him in a treatment program. As the man, an agnostic, lay on his bed one afternoon, he decided to play a thought game: If there were a God, what would God look like? He closed his eyes and imagined a "blue-white star." He reached out, snatched the star, and touched his hand to his chest.

"As soon as I put it into my chest, something took over my body," he told Miller. "It was physical. I could feel my skin bulging outward, and I started to gasp for breath. I felt an ecstasy that was—the only way I can describe it is that it was just like a sexual climax, except that there was nothing physical to it and it was better than sexual climax, infinitely better.[21] This thing had control of me; I mean, something took over my body. I was not in control. Something was doing this to me. . . . When it left me, I just wept. I was just stunned. I thought, 'What the hell was that?' I mean, it was real. And I was saying, 'Thank you, thank you, thank you, my God, thank you.' "[22] From that moment, he stopped drinking, and he quit his high-stress job.

In his book, Miller picked up William James's gauntlet from nearly a century earlier, and he found the very same elements that James identified: union with the universe, the peace and love, the feeling that something external was acting on them, and the conviction that this experience was more real than everyday life. And, like William James, Miller accepted these stories at face value, not cramming them into a file labeled "Psychosis" or "Sexual Disorder" or "Head Injury." Admiring his pluck, I set out to do the same. As a reporter, I try not to rely on other people's research. I wanted to gather my own stories and run them through the filter of my own close questioning. And I suspect because I had experienced a little of the mystical myself, I wanted some reassurance. If these people weren't crazy, neither was I.

A Nation of Believers

Over a three-month period, I interviewed more than two dozen people who testified that they briefly brushed against the fringes of another, nonmaterial, dimension, and were transformed by the contact. Each time I finished an interview and turned off the tape recorder, I tried to identify how I felt about these modern-day mystics, like Sophy Burnham, who saw "the hem of God's garment." Intrigued? Yes. Awed? Certainly. But in the main, I felt a little deflated. Why were they chosen to enjoy this overpowering event, and not me? I couldn't help suspecting that God might be an arbitrary schoolmaster, dispensing gold stars to the talented but naughty (think Mozart) and withholding them from the earnest but dull (think Salieri).

Of course, most people in the United States have never felt themselves dissolve into union with God. But according to surveys,[23] fully half of Americans claim to have experienced a life-altering spiritual event that they could circle on the calendar in red ink.

It appears that spiritual *experience* (as opposed to *belief* in the tenets of a religion like Catholicism or Islam) occurs as often today as it did a century ago. Even the twentieth century, with its Freuds and B. F. Skinners, its technological advances and scientific reductionism, could not quash Americans' yearning for the divine.

On the theory that the extreme cases paint the spiritual realm more vividly than moderate ones, I did not seek out people like me, who had experienced God in a brief religious shiver. Rather, I sought out people who had traveled to a different spiritual continent and back, and were willing to tell their stories.

Finding a mystic, it turns out, is pretty simple. They are everywhere. Many do not broadcast their experiences, worried that they might be considered odd. I located some of my mystics by asking friends, or by calling up people I had interviewed while on the religion beat at National Public Radio. Then I stumbled onto a mother lode of mystics: I called up Cassi Vieten at the Institute of Noetic Sciences, a group that studies the intersection of science and spirituality. IONS had just conducted a survey of people who had undergone a dramatic spiritual transformation. Cassi sent an e-mail to four hundred of them, describing my interest in their stories and giving them my e-mail address. Within a week, more than eighty people had sent me long essays about their experiences, which were often—literally—incredible.

Since I could neither confirm nor debunk their stories, I asked myself two threshold questions to determine whom to include in my research. First, is this someone I would want to join my friends and me for dinner—that is, is he or she educated, reasonable, accustomed to appearing in public fully dressed? And second, did this person wonder about his sanity after his epiphany? (The answer to this question had to be yes, as that would indicate he or she had not lost touch with reality.)

As I heard their stories, I concluded that three elements are essential to profound spiritual experience, and all of them echo William James's insights. First, the moment itself is weirder and *more real* than everyday reality. Second, it gives the person radical new insights about the nature of reality and the nature of God (the noetic quality). And third, those insights appear to rewire the person and his approach to life.

A Different Sort of Reality

I met Arjun Patel during a tropical storm in Miami, on December 6, 2006.[24] I had barely made it to the hospital lobby where he worked when the heavens opened, the rain slamming down in angry sheets.

Soon a crisply dressed man in his mid-thirties burst through the doors, splattering raindrops in an arc around him. Enter the hospital's smart, young psychologist who specialized in helping patients with terminal diseases (and their families) negotiate the final yards into death.

We sat down in a sterile examination room. I gazed at Arjun, dressed in a button-down blue shirt, fresh and unwrinkled beneath his gray-and-blue-striped tie. He had beautiful olive skin, a near-shaven head, and a goatee, and he radiated a serenity that surely calms his patients in their final days.

Raised in a Hindu home, Arjun had experimented with meditation in his teens. But in his freshman year at the University of California, Santa Cruz, he and a friend decided to meditate every afternoon at five p.m. After two weeks, Arjun said, he started to "play around a bit" and silently chant a Tibetan Buddhist mantra. He remembered the scene with cinematic precision. He and his friend were sitting cross-legged in their tiny dorm room on Wednesday evening in February 1990, and as the long shadows of twilight crept across the room, Arjun closed his eyes and slowed into a rhythmic breathing.

"I would sometimes see things when I'm meditating, just natural, normal distractions, and little visual things that happen," he explained. "But this was different. It was dark, and there was this little pinpoint of light that just kept moving closer, relentlessly closer. I thought, *What is this?* It didn't look the way I imagined any mystical experience to be, and I wasn't really looking for one. So this light was off in the distance and it kept getting closer and closer, and there was almost a sound that came with it, like a roar."

I recalled Sophy Burnham's words—*"there was a hollow darkness . . . sort of like an oncoming train."*

"I was trembling. I was vibrating, I had a tingling in my fingers," Arjun continued. "Every time I talk about it I feel it again, a kind of clamminess in my hands. I felt very, very weightless, very light and easy. My breathing was very shallow, and I think I was exhaling really hard, and I was rocking back and forth. I could tell I was smiling from ear to ear. At some point, as the light got closer, there were two eyes looking back at me. They were stylized eyes. I later learned they're called 'Buddha's eyes.' And they kept moving closer and closer, and the eyes were half opened and half closed. And when it got up to where I was, the eyes blinked. And it was like . . . I don't really know how to describe it. The light wasn't out there anymore. I wasn't in here anymore. Everything just sort of collapsed, and there was nothing left but light."

"Were you light itself, or was it in you?" I asked.

"There wasn't any 'me' anymore. There was no boundary, no 'This is me and this is something else.' There was just total seamlessness."

"Did you feel a loss of identity?"

"Yes, absolutely."

"And did that feel good or bad?"

"It just felt like, *This is the way it's supposed to be.*"

Arjun slept little over the next two weeks. He listened to music and kept to himself. And he felt a surreal unity with all things.

"I remember lying on the grass under the tree in the quad," he said. "I remember just running my hands through the grass, and just feeling for the grass."

"Feeling for the grass," I repeated. "You mean, empathizing with it?"

"Yeah. It was alive, the grass was alive, and it was beautiful. There was no self-reference point."

The euphoria eventually ebbed, although, Arjun said, every time he talks about it "my body remembers what it feels like." Sixteen years passed. When I met him, Arjun was meditating twice daily, and considered that moment the most important of his life. Since then, he has married, started a family, and, as an end-of-life psychologist, has eased hundreds of people through sickness and death.

I told him I did not know how he coped with the emotional drain of his job.

"It was the light," Arjun said, and laughed self-consciously. "I don't ever, ever, *ever* talk about this with my patients. But, you know, I feel that the experience—the light, whatever you want to call it—it's always there. It's kind of like a well: you can dip down into it and draw some of it up when you need it for spiritual energy and emotional energy. It's something you always have access to."

He paused.

"It was an eternal moment. It doesn't matter that it was sixteen years ago. It's happening now."

A Different Sort of God

One theme kept drawing my attention, like a buoy that repeatedly popped up in the choppy sea of research. No matter what their belief systems, the people who had enjoyed profound spiritual experiences saw the same kind of "Other." In fact, they often shunned the word "God," since God is the purview of particular religions. Without exception, this being in the eye of their spiritual tornadoes was a loving presence, infinitely intelligent and gentle. My mystics might use different terminology, but they described uncannily similar experiences: love, peace, and, often, an overwhelming sense of unity with the universe—and always, *light*.

A Sufi mystic named Llewellyn Vaughan-Lee, for example, described his first mystical experience as a light switch flipping on.

"I saw light dancing around everything. You know when you get sunlight sparkling on water? It's like that. I felt really alive with this quality of joy and laughter. And I just knew, *this* is the way it is, and everything else is the way it isn't. Everything else is covered up. Everything else is veiled."

Undergirding the light was love, Llewellyn said, a physical love, the kind that Saint Teresa of Ávila experienced in what can only be called spiritual orgasms.

"You long for God," Llewellyn explained, "and if the spiritual love is awakened, that draws your attention closer and closer and closer to God, until finally you have these mystical experiences of union, in which the ego falls away and you are drawn into the ocean of love. There is no 'you' anymore, there is just divine love, which is what the mystic longs for."

So it was for Adam Zaremberg, a young Catholic man who had recently converted from Judaism. His first encounter with God was a visceral event. It was like downshifting from third gear to first: the world bucked with pent-up energy. One spring day in 1999, when he was a senior in high school, he spotted an underlying reality, and in the blink of an eye, the mundane was teeming with explosive energy

"I just sat down on a bench, and suddenly I was able to see the world in a new way. Everything had just changed. It was like... all of creation emanating itself. It was just infinitely more itself. Like everything seemed magical and filled with energy... but it didn't change the way anything looked. I came up with an analogy for it: it was like a grape squeezed to the popping brink."

"So it's still in its same form," I clarified, "but it's bursting...."

"It's bursting at the same time. It really is radical—everything changes. And it was accompanied by a very calm feeling, where you're glad to be there, glad to be alive, not afraid of anything."

A Buddhist, a Muslim, and a Catholic have similar mystical experiences—it sounds like a bad joke, but it is cosmically accurate. I asked Bill Miller at the University of New Mexico if he had found any particular themes in the accounts of his "quantum changers."

"All the people who described the mystical type of experience described the same 'Other,'" said Miller, who is a Presbyterian. "Never mind their religious upbringing—they were in the presence of the same thing."

"What is that 'Other'?" I asked.

"It's profoundly loving and accepting beyond words. It doesn't demand anything of you at the moment. You simply feel accepted as you are, and loved as you are, to the very depth of your being."[25]

Many of these people did view this "Other" through the lens of their faith. In their everyday spiritual practices, Arjun Patel, a Buddhist, saw Buddha's eyes; Llewellyn Vaughan-Lee, a Sufi, communed with the Beloved; and Adam Zaremberg, a Catholic, visualized Christ. But none of them claimed his "God" as the only authentic God, nor did they begrudge anyone else a different view. They were like witnesses to the same God but from different angles.

So prevalent was this nonreligious "Other" that I wondered why these people would remain content with a particular religion. I mused about this with Sophy Burnham, who had spent years delving into Asian religions and ended up where she started, in the Episcopal Church. I told her I was surprised by that.

"So was I," Sophy said.

She explained that she thinks of spirituality as a wheel with spokes leading to the hub. "And each of the spokes is a path to God—is a religion, if you will, or a spiritual practice. And they will all take you to the direct experience of God. But you have to choose one and go all the way down it. If you dive down one and test it for a minute and you come back out and go along the circumference again, and dive down another, you'll never reach the hub."

I must admit, I disliked the analogy. If truth be told, all this interfaith Kumbaya unnerved me. A decade earlier, when I had been spiritually adrift, it wasn't a nondescript "Other" who slipped the rudder into place and breathed life into my faith. It was Christ who met me. Moreover, I *liked* the idea of God's becoming a person and going to weddings. I *liked* the man who outargued the lawyers, who outpreached the rabbis, who declined the help of the governor of Jerusalem to stave off His own wrongful death, yet spared the life of a prostitute on the verge of legal execution.

And what was I to do with Jesus' words, "I am the way, the truth and the life. No man cometh unto the Father, but by me"?[26] Jesus did not fudge His words. His vision of the "Other" is no fuzzy thing, but a God with likes and dislikes, who has a personality and a plan.

And yet, I had always squirmed when my fellow believers declared there is but one way to God. I had chafed at the assertion that a philanderer or an embezzler or a rapist who asked Jesus into his life would take the express train to heaven, while Mahatma Gandhi writhed in hell because he didn't. For that matter, I have never been comfortable with heaven and hell, much less certain about their residents. The stories of my modern-day mystics were like a crowbar prying open my faith. I had opened a Pandora's box.

The Cost of Transformation

These spiritual experiences, it seems to me, have the air of high-priced interior decorators on reality television shows: they arrive (unbidden), and immediately begin punching through walls and removing countertops. They haul away most of the furniture, including your favorite red leather chair, and what they deign to leave, they put in the basement. These experiences don't ask, they simply move in, insisting you'll *love* it when they're done, and then with a brisk clap of the hands, they're gone, leaving you stunned, panting, and utterly transformed.

I could hear all of that in Susan Garren's voice.

In the summer of 2004, Susan was thirty-seven years old and living in Asheville, North Carolina. She had started her own real estate appraisal business, which was flourishing, and she had begun dating a handsome young doctor named Vince Gilmer. On July 5, the police arrested her beau for the murder of his father. (He was later convicted of first-degree murder.)

Susan considered herself self-reliant: she managed her business and owned a house. She ran in a sophisticated circle of friends. A nervous breakdown was not in her makeup. But after Gilmer's arrest, she retreated to her bed, traumatized.

As she lay there, she said, one by one, her senses ceased to operate.

"All of a sudden I realized I wasn't able to see anymore. Then I couldn't hear anything. But I could *sense.* And I heard a voice—well, I didn't technically hear it. It was some sort of communication that I could understand. And it was just reassuring me that everything was okay. I felt very calm, very safe. The more I relaxed into it, the safer I felt."

"How would you define this voice or this experience?" I interrupted. "Is it God? Is it the universe? Is there a personality to it? Is it infinite mind?"

"I call it the Source," she said. "It's the Source of all that is, and everything that encompasses it. It wasn't the old man with a beard; it was some very powerful force able to communicate in extrasensory ways. But it was very loving, very comforting. Every time I think about it I begin to tear up. It's absolutely the most powerful feeling I've ever experienced."

"How did it end?"

"I came out the same way that I went into it. The first thing I could remember is I could hear the birds chirping in my yard. And then I was back in this consciousness, and I lay there for however long, thinking, *Holy shit, what just happened?*"

She laughed, and then her tone grew serious. "I've changed on a cellular level," she said. "I literally feel that I'm being rewired. I used to be quite the party girl. I was out all the time. I used to date for sport. And since that happened, I suddenly had no desire to live a superficial lifestyle. It was night and day—this sudden, absolute, quick change. Now a great day for me is playing in the yard, or having a great conversation with my friends at my house, or reading a great book. It *completely* changed my lifestyle."

Susan had always thought of herself as a spiritual seeker, interested in alternate beliefs. But she never pursued them much past flipping through a New Age magazine. Since that moment, however, she has been obsessed with quantum physics. "Where I thought about it before, now I'm pretty much consumed with it."

"Why does quantum physics interest you?" I asked.

"Quantum physics tells me that we're much more connected than we realize." I thought of Arjun's connection with the grass.

"What did your friends think of these changes?"

"Well," she drawled, "the friends I've known for fifteen, twenty years enjoy this deeper aspect. People who knew me only four or five years I don't have as friends anymore. I have very few friends, and that is intentional. I just weed them out. I can sense people who have no depth. I don't mean to sound rude, but after I changed, I wanted to be around others who had deeper philosophies, who were interested in exploring spirituality, who were interested in bettering their lives and empowering themselves. And if people weren't interested in that, I didn't want to be around it."

Just like Sophy Burnham. And just like me.

The First Became Last and the Last Became First

When I told Bill Miller at the University of New Mexico about my conversations with modern-day mystics, he merely nodded.

"It's a one-way door," he said. "It's not like you *decide* not to go back" to your previous lifestyle and priorities. "The experience people describe is: I just *am* different."

"How did your subjects' values change?" I asked, referring to the people he interviewed for his book *Quantum Change*.

"They were turned upside down," he said.

Miller explained that he had asked the fifty-five people in his study to look at a list of fifty values, and rank them according to what was most important before and after the mystical experience.

"Essentially the things that were at the top of the hierarchy [before the experience] went to the bottom," he said. "Often what was literally number one was number fifty, and vice versa: the first became last and the last became first."

Before the experience, men ranked their top personal values as: wealth, adventure, achievement, pleasure, and being respected (in that order). After the experience, their top values were: spirituality, personal peace, family, God's will, and honesty.

The women seemed to have fewer self-centered values than the men to start with, but even these shifted: from family, independence, career, fitting in, and attractiveness (before the mystical experience), to growth, self-esteem, spirituality, happiness, and generosity (afterward).

Often these subterranean changes flowered into a new career or life course. Llewellyn Vaughan-Lee became a Sufi mystic and author, for example, and Arjun Patel chose to counsel the dying because of the "light."

Sometimes these changes dislocated their lives. Usually, the transformed people felt a twinge of regret at losing their former life but found invariably that the spiritual adventure more than compensated. I felt sorrier for their family and friends, who became the "collateral damage" of the spiritual experience, scratching their heads helplessly as the person they thought they knew disappeared forever. Virtually every woman I interviewed, and several of the men, reported that their values and goals had veered so drastically away from their spouses' that they eventually divorced. Asked why she and her husband (whom she still loves) parted ways, Sophy Burnham replied, "I wasn't the person that he had married."

"Have you paid a price?" I asked Llewellyn, the Sufi mystic.

He chuckled wryly. "Ah, dear, yes. There's always a price. The price is yourself."

"Is it painful?"

"There's nothing more painful. You become incredibly vulnerable, you become incredibly naked. Nobody in their right mind would want to do it."

"Except you can't help it."

"There you go," Llewellyn said. "You don't have a choice."

What I did not realize when I said good-bye to Llewellyn was that the stories of my mystics would give me a peek into the rest of my research. Their stories would contain elements that would prove central to an array of different spiritual experiences: the sense of union with all things and the universe; the loss of fear of death; the new definition of reality and of "God"; and profound personal transformation. I heard some or all of these descriptions from people who experienced emotional breakdown or mental dysfunction, who experimented with psychedelic drugs or meditation, who had near-death experiences. I didn't know it then, but I was on my way to redefining for myself the nature of God and reality.

But I am getting ahead of myself.

Fortified by the insight that spiritual experience is not only everywhere but open to research, I was ready to tackle the questions that were driving my own quest. I began by reexamining the faith of my childhood in the light of science.

CHAPTER 3

The Biology of Belief

MARY ANN BRADLEY SAT PERCHED on the living room sofa in her Nantucket home. Her eighty-five years had been kind to her. She always had been short, but these days barely topped five feet, with a perfect mane of silver hair and not a single wrinkle on her forehead. It was August 18, 2006, about ten in the morning. She had prepared for this moment, as evidenced by several pages of handwritten notes spread out on the coffee table before her. She was, and is, one of the dearest people in my life. And at this moment, Mom was serving as my guide to a religion that is, perhaps, a hundred years ahead of modern science—a religion that relies wholly on the power of thought to alter the body.

"I have never known anything but Christian Science," she began. "It's been the guidance system of my life and has never let me down."

Mom comes from a family of Christian Scientists, and raised my brother and me in the faith. Her mother was a "practitioner," the metaphysical equivalent of an emergency-room doctor. People called her with their problems. Granny prayed for them, and more often than not, a healing ensued. What most people know about Christian Scientists is that they do not take medicine—even vitamins—and that they rarely go to doctors. What most people do not know is that there is a method to this asceticism.

Christian Science holds as a central premise that healing is a function of spiritual understanding; that matter and its conditions, including sin and disease, are "false beliefs"; and that prayer changes a person's thought, which results in healing. An example drilled into me by my Sunday-school teachers: When a person looks at a dirty bathroom mirror, his reflection may be marred by smudge marks and toothpaste splatters; but the problem is with the mirror (the distorted image of reality) and not the person. In Christian Science, the way you clean the mirror—and restore the reflection—is to clean up your thinking.

"Everything is thought in Christian Science," Mom explained. "Everything is going on in your thinking. If you remove the 'false beliefs' which we can call error, evil, sickness, and replace those thoughts with the spiritual truth—the truth that man is made in the image and likeness of God—then the body responds. Since there is nothing broken in God, then there can be nothing broken in the image and likeness of God, man."

Okay, let me translate for those unfamiliar with the Victorian language of Christian Science, which was founded by Mary Baker Eddy in the mid–nineteenth century. Christian Science prayer shares little with the popular conception of prayer. Rather than beseech an authoritative and exceedingly busy Judge to stop what He's doing, listen to the plea, weigh the arguments, and then miraculously intervene, Christian Science appeals to spiritual principles, akin to working out a math problem. In this sense, Christian Science prayer is nearer to meditation than to petition: it is a mental discipline, one that claims that when you apply spiritual "laws," you take control of your environment—your body, your employment status, your love life, your mood. Which does not mean that life proceeds perfectly or endlessly—not even Mary Baker Eddy cheated death—but that what you think directs how you experience the world. Or as my mom always says, "Your thinking is your experience."

I asked Mom for a concrete example.

"Okay, I'll tell you about my broken hand," she offered. "I was engaged to be married and it was a Sunday afternoon. I was with my fiancé—your dad—and he was teasing me, and I hit his knee in response. I knew immediately something was wrong with my hand. So I asked him to take me home.

"I started working in Christian Science," she continued. "And this is the thought I worked with: Nothing can come into my human experience that I do not allow in my consciousness as a reality. And it was up to me to see that this broken bone—or whatever it was, at that point I didn't know—was unreal. Just as the statement 'Two plus two is five' is erroneous, unreal. You take it out of your consciousness as a reality. You erase it, and then you substitute the truth."

"In this case, what was the truth?" I asked.

"The truth was, that I could not have anything broken," she said. "I was a spiritual idea of God, so there was nothing to be broken."

Then Mom grew quiet, overwhelmed by the memory of what came next. I had seen this cascade of emotion before, when I talked with Sophy Burnham, and Arjun Patel, and virtually all of the mystics I had interviewed.

"I was sitting on my bed, about to go to work," she recalled. She was reading the Christian Science textbook *Science and Health* when the tectonic plates of her reality shifted.

"I had this great sense of light—of one thing flowing out of another, out of another, out of another, out of another, into eternity," she said. "There was nothing but light. It's all one. It's all God. The all-ness of God, which is the oneness of God, and I was within that oneness. And I just sat there, and it didn't last very long, but I'm quite convinced that that was when my hand was healed."

At the request of my father, who had not yet converted to Christian Science, my mother visited a doctor who X-rayed her hand two days later.

"And the doctor came back and said, 'Well, yes, the large bone in your right hand is broken,'" Mom recalled. "And he said, 'What we usually do in cases like this when someone hasn't gone to the doctor immediately is we rebreak it and reset it. But if you want that done, you're going to have to go to another doctor. Because it's perfectly set and it's almost healed.'"

Mom paused, reliving the intensity of that moment. "And I remember walking out of his office—my feet didn't touch the ground, I was so filled with God's truth, the spirituality, the marvel of it. That was the end of it."

As I began to study the biology of belief, I found myself circling back to Christian Science and taking a fresh look at healing—what believers see as the evidence of divine laws in operation. Certainly I continued to delight in swallowing an aspirin or cough medicine anytime I chose. But the more I talked to people about spirit and matter, the more I suspected that Christian Science was onto something.

Laughing Back to Health

Mind-body medicine has become so widely accepted today that it is difficult to recall when it was considered fantasy. For ordinary Americans, the conviction that your thoughts or emotions affect your body gained traction in the 1950s when Protestant preacher Norman Vincent Peale wrote his transformative book, *The Power of Positive Thinking.* But it was not until the 1970s that *scientists* finally began to acknowledge a connection between mind and body.

Anne Harrington, a professor of the history of science at Harvard University, identifies Norman Cousins as the man who triggered the revolution. In the 1970s, Cousins, an influential writer and editor in chief of *Saturday Review*, was hospitalized with a mysterious, crippling disease. Doctors diagnosed some form of progressive paralysis or a degenerative bone disease that would eventually kill him. Facing a death sentence, Cousins threw out the specialists, checked himself out of the hospital, and worked with his own physician to wage a novel war on the disease. His plan was the medical equivalent of unleashing millions of beagles in Baghdad: he flooded the place with good cheer.

"He knew that there was research and evidence showing that negative emotion—fear, anger, anxiety—was bad for you," Harrington said. "But he felt that there had been little study of whether *positive* emotions might have the *opposite* effect on your health, that it might be good for you. He felt he had nothing to lose, because he wasn't going to get better through conventional means, and perhaps he had a lot to gain.

"So he checked himself into a hotel," she continued. "He had films of *Candid Camera* and the Marx Brothers brought in. He read all sorts of funny books, and he discovered that ten minutes of a belly laugh gave him twenty minutes of pain-free sleep. And little by little, as it came to be famously remembered, he laughed himself back to health."[1]

What earned Cousins a place in medical history, Harrington said, was the fact that his experience was published in the *New England Journal of Medicine*, one of the world's premier medical journals. Cousins received close to 3,000 letters from doctors and researchers, who praised him for medically documenting his novel approach. Firmly clutching the gauntlet thrown down by a mere layman, scientists began to explore how Cousins's recovery could be explained within the parameters of science.

Thus was born a new science: psychoneuroimmunology. The infelicitous name makes sense when you break it down: Your thoughts and feelings (psycho) affect the chemicals in your brain (neuro), which affect the hormones that fight disease or replicate viruses (immunology).

Psychoneuroimmunology. New research centers began to spring up—at Harvard, Ohio State, the University of Rochester, and the University of Miami, and one named after the man himself, the Norman Cousins Center for Psychoneuroimmunology at UCLA.

The research flowed quickly, and showed that nonphysical things like thoughts and emotions affect our bodies at the cellular level, just as surely as do genes or lifestyle or the medicines we take. Emotions—particularly depression and stress—are linked to heart attacks.[2] They suppress the immune system as it tries to fight the flu.[3] One's thoughts and attitudes affect the course of cancer,[4] and the recovery from breast cancer.[5] Emotions even affect how long one is plagued by the skin condition psoriasis.[6]

As I investigated these findings, I stumbled upon a researcher who has found evidence that *spiritual* thinking may be the most powerful mental antidote of all. I flew down to meet Dr. Gail Ironson at the University of Miami and a patient of hers whose spirituality seems to have kept AIDS at bay.

God and HIV

"It was March 4, 1994. About two-twenty p.m.," Sheri Kaplan recalled. She smiled wanly. "You remember those things."

A few days before she lost the illusion that life is long, Sheri had walked into a health clinic near her home in Miami. She was in her late twenties, embarking on a new romance, chuffed with life.

"I said, 'Give me every test. I'm going to do a complete physical checkup so I can feel good about myself.' "

Sheri strode into her doctor's office a few days later. The doctor looked down at the table, reluctant to meet Sheri's eyes.

"I knew. You can feel the coldness in the air as soon as you walk in. It was fear in the air. And when I sat down and she told me that my test results came out positive, I said, '*Noooo.* What are you talking about?' And she goes, 'Yes, Sheri, the test results came out positive. We checked and then double-checked. It is confirmative—you have HIV.' And I don't know what else she said after that. Her lips moved, but nothing came out. I could have cut my fingers off and wouldn't have known it, I was so numb."

Her questions waged a rapid-fire assault: "Who did this to me? How long have I had it? Will I be able to keep my job? Will I get married? How am I going to tell my mother? My father? Will people point fingers at me? I'm twenty-nine years old—*how long am I going to live?*"

When I met Sheri in December 2006, more than a dozen years had passed since that moment. A slim forty-two-year-old, she had wavy red hair and a tanned, freckled face. She wore no makeup, and didn't need to, with her intense blue eyes and wide smile.

She chatted unself-consciously as a researcher at a University of Miami clinic drew several vials of blood. They compared notes on CD4 cell counts, viral load, and a battery of other measures that were the lifeblood of those initiated into this dreaded disease. Every six months for nearly a decade, Sheri has dropped by the clinic to meet with Gail Ironson, a doctor and professor of psychology and psychiatry. Twelve years after her diagnosis, Sheri has somehow beaten the virus into submission. She remains healthy and has never taken so much as a pill for medication. The question that drove Gail Ironson was: *Why?* Why do some people with HIV never get sick?

Sheri's personality holds some clues. She's pugnacious. She refused to surrender to the disease. Instead she opened a nonprofit support center—"where people can laugh, and fall in love, and meet people, and have a life after HIV." Under her determined management, the Center for Positive Connections expanded from a nine-by-twelve-foot room to a 3,500-square-foot facility with a half-million-dollar budget and nine employees.

But something more was at work than a renewed purpose to life. An indefinable current, a *pneuma,* gave lift to her goals. Sheri's was not a traditional God—she was raised Jewish but believes in reincarnation—but a living, breathing entity who served as a copilot to steer when she could not see straight for the terror.

"At first I thought, How could God do this to me, make me the leper of society?" she said. "Then I realized, I was chosen. The message was, I was chosen so I could help create social change, so I took this as my role. I realized God didn't want me to die, or even get sick."

Gail Ironson has been studying people like Sheri Kaplan for more than a decade, looking for clues to longevity. Her hunt began in the mid-1990s, when Ironson, who trained at Stanford and the universities of Miami and Wisconsin, launched a longitudinal study of people living with HIV. She noticed that a rare group of people fared much better than others. Back then, before the most effective drugs were developed, the average life expectancy for those diagnosed with AIDS by opportunistic infection was a year and a half.

"But many of our people were alive seven, eight years past diagnosis," Ironson said. "And we were looking for both psychological and immune factors that might be protecting their health. And we found both."

The long-term survivors were less depressed. They were better at coping. They were more proactive, finding the best doctors and the best research. Ironson could have predicted all these characteristics: studies have long indicated that your attitude—whether you curl up on the ropes or deliver a counterpunch to the kidneys—often affects the course of a disease. But Ironson noticed another, unexpected trait among her long-term survivors.

"People kept talking about how spirituality was important in their lives," she said. "If you ask people what's kept you going so long, what keeps you healthy, often people will say 'spirituality,' or 'my relationship with God.' I mean, there were many other things that came up, but spirituality came up a lot as a moving force in people's lives."

Ironson began to zero in on spirituality as a predictor of how fast the disease would progress. First, she compared those patients who felt embraced by God or abandoned by God, and tracked two biological measures—CD4 cells and viral load. CD4 cells are a part of the immune system that helps fight off tumors and viruses such as HIV. The HIV virus also attacks CD4 cells, so as the disease makes inroads, it knocks off more and more of these fighter cells. Viral load is a measure of how much of the HIV virus is in one's system; the higher the viral load, the sicker the person becomes.

Ironson found that people who felt abandoned by God after their diagnosis lost their CD4 cells at a rate 4.5 times faster than the people who relied on God to cope with the diagnosis; their viral load also increased.[7]

Let me repeat that astonishing statistic: *Turning to God rather than rejecting God appears to boost your immune system and stave off the disease nearly five times as effectively.*

Next, Ironson looked at people's "view of God." Do you believe God loves you or that God is punishing you? She found that people who held a positive view of God maintained those CD4 cells twice as long as those who did not. And people who believed God loves them maintained the cells three times as long as those who felt God did not love them.[8]

Ironson compared one's spirituality and view of God with a battery of other items that affect the course of the disease, including the mother of all predictors, depression. Depressed people saw their CD4 cells disappear twice as fast as those who were not depressed. But if they embraced a spiritual outlook on life, that more than offset the bad immunological effects of depression.

"I find that extraordinary because depression is the most consistent, reliable predictor of how people do with illnesses, and not just HIV," Ironson told me. "People who are depressed are much more likely to suffer a second heart attack or die. People diagnosed with cancer who are depressed are much more likely to have a recurrence and have a poorer disease course. So depression is a very well-established psychological factor, and to find another psychological factor that may potentially be *more important* is very surprising."

Ironson and her researchers looked for alternative explanations as to why spirituality might be related to better health. They ruled out other explanations through statistical analysis. They controlled for age, education, gender, and race—even church attendance, which has been linked to longevity. They controlled for psychological variables such as optimism, life stress, depression, and coping.

"We wanted to determine whether there is some independent contribution of spirituality *over and above* other psychological constructs that other people had looked at," she said.

"And what did you find?" I asked.

"We found that spirituality still predicted a significant amount in disease progression."

"Just so I understand it," I said, "you found that if, say, someone wasn't taking their meds and was depressed, they could still do better if they were spiritual than if they weren't spiritual?"

"Yes. Now, I'm not in any way suggesting that people shouldn't take their meds," she added, laughing. "This is really an important point. However, the effects of spirituality are over and above. So whether people are taking their meds or not, spirituality was still related to slower disease progression."

She paused a beat, to let the idea sink in. "Spirituality is our most powerful predictor to date."

While scientists might balk at the idea that "God" has anything to do with HIV progression, they readily agree that your *thoughts* affect your body. Here's how it might work with HIV. Stress hormones that make your heart race and hands sweat, such as cortisol and norepinephrine, accelerate how quickly the HIV virus can replicate. Ironson said her research has shown that the stress hormone cortisol is lower in people who score high on having a "sense of peace" through meditation, belief in God, and other spiritual practices. She has found that norepinephrine is lower in people who score high on "altruism" and compassion, another component of spirituality.[9] In other words, Ironson connected the dots from a patient's spiritual beliefs, to the chemicals in her brain, to the immune system.

Her patient Sheri Kaplan put it this way: "I have the power to control my mind. That's one thing that I can control, before it gets to a physical level or an emotional level. So if you can nip it in the bud, you can stop anything from moving forward into a direction you don't want it to. If I visualize the virus not being there, the virus is not there."

"You're saying you can overpower the virus with love or good thoughts like washing it out of your body?" I asked her.

"Oh, yeah. Washing it out—I do that every day in the shower. I watch the virus go down the drain."

Scientists would differ with that conclusion. The virus is still lurking in Sheri's body. But tell that to the HIV, which has failed to make an inch of progress since the day it invaded her life.

Is Anyone Up There?

Neither Sheri Kaplan's story nor Gail Ironson's research claims that there is a God who puts a restraining hand on the HIV. It is the *belief* that there is a God who guides, not abandons, loves, not punishes, and occasionally intervenes to cause the miracle. This keeps everything in a closed—and safe—loop. The materialist can argue that the power to affect one's own body originates in the mind of the believer, not from an external or supernatural source. In other words, the skeptic can point to a material mechanism, and there's no need for a God to fill in the gaps.

So now let's launch into far more turbulent scientific waters: the prayer studies. The premise of these studies is that one person's thoughts (or prayers) can affect another person's body. The vast majority of Americans believe in the power of prayer, and spend a lot of time demonstrating that belief, whether in church, or the hospital room, or as part of their morning devotion. And why not? It's all mind over matter anyway, right?

Apparently not. I soon learned that conflating prayer (for someone else) with the mind-body connection amounts to scientific blasphemy. Prayer and positive thinking may *appear* to share some characteristics, scientists told me, but that is a mirage. It is a little like equating the opening scene of the movie *Saving Private Ryan* with actually losing your leg on Omaha Beach in the D-Day invasion. They seem to portray the same sort of event, but in reality everything is different, including the mechanism by which they are experienced—pixels of light on a movie screen versus a physical bullet shattering your thighbone.

I asked Anne Harrington at Harvard to explain why some scientists embrace the mind-body connection in one breath and repudiate intercessory prayer in the next.

"It's a gigantic leap. It's a whole different ball game," Harrington said.

When talking about the mind-body connection, science can explain the mechanism in which a person alters his own experiences through prayer and mental discipline.

"But with interpersonal prayer, you're making, at least potentially, a metaphysical claim—a claim about the nature of external reality. You're arguing either for the existence of invisible mysterious forces—which science doesn't know anything about—which somehow emanate from the praying person to the person being prayed for. Or you're making an argument for a miracle, for an intervention into a disease process that would not have happened otherwise. I think it's a very big step."

One of the first scientists who dared to test God (and the wrath of his colleagues) was R. C. Byrd at San Francisco General Medical Center. In the late 1980s, he monitored nearly four hundred patients admitted to the coronary care unit for heart problems. Half the patients received prayer from Christian intercessors, the other half received no prayer. The patients knew they were in the study (they signed consent forms) but no one—not the patients, not the researchers—knew who was receiving prayer and who was not. This eliminated the placebo effect, in which a person who thinks he is receiving a healing agent (a drug, or, in this case, prayer) actually improves—not because the drug (or prayer) is effective, but because the patient *believes* it is.

After ten months, Byrd's study seemed to indicate a medical impossibility: those who received prayer had many fewer hospital stays and much less need of medical attention, including ventilatory assistance, antibiotics, and diuretics.[10]

The study rocked the materialists back on their heels, and emboldened other researchers to assess whether there exists some nonmaterial force that responds to prayer and heals. One study found that prayer physically helped people with advanced AIDS.[11] Another large study replicated the Byrd findings and found that prayer helps one recover from heart attacks and heart disease.[12] There was good news for would-be parents: in a controversial study, researchers in South Korea found that women who were trying to become pregnant through in vitro fertilization were twice as likely to conceive if they received prayer than if they did not.[13] Monkeys, too, have much to celebrate: those who received prayer healed more quickly from wounds than those who received no prayer.[14] And, finally, in a study destined for the annals of the weird, Israeli doctors tested *retroactive* prayer. They asked intercessors to pray for half of nearly 3,400 people who suffered bloodstream infection in the hospital in the early 1990s. But the intercessors began praying in 2000—six to ten years *after* the patients developed the infection. The group receiving retroactive prayer had lower mortality rates, shorter hospital stays, and shorter periods of fever.[15]

But for every study suggesting that prayer heals a person's body, there is another one showing that prayer has no effect—or worse, that you don't want prayer, no how, no way, *get that intercessor away from me.*

Does prayer help people with heart problems in a coronary care unit? Researchers at the Mayo Clinic found no effect.[16] Does it benefit people who needed to clear their arteries using angioplasty? Not according to researchers at Duke and other medical centers.[17] In other studies, prayer and "distant healing" did not seem to affect the course of HIV,[18] did not alleviate pain for those suffering from rheumatoid arthritis,[19] and did not ease the plight of those on kidney dialysis machines.[20] People with skin warts will get no comfort here: researchers found that people who received prayer saw the number of warts actually increase slightly, compared with those who received no prayer.[21] In the most famous study, conducted by Harvard researcher Herbert Benson and his colleagues, prayer backfired, at least for those who knew it was coming.[22]

I looked a little closer at this study, which involved 1,802 patients undergoing cardiac bypass surgery. They were split into three groups. Members of one group received intercessory prayer for fourteen days, beginning the night before the operation, but they did not know it. Members of another group were told they would receive prayer, and they did. The third group did not receive prayer. The result left advocates of prayer sputtering and critics of prayer chortling. Patients who received prayer and did not know it, and those who received no prayer fared about the same. The patients who did worst—who had the most complications—were those who received prayer and *knew* it.

I have heard many explanations for this surprising result, including the complaint that these prayer studies do not reflect the way people actually pray. How many of us pray for a complete stranger by reading from a script? Most of us pray urgently, passionately, for a person we know and love. It is an intimate act. This critique comforts some people as far as the results of studies showing that prayer has no effect. But what about the Benson study showing prayer is *bad* for your health?

One explanation, Anne Harrington told me, is the potential fear that arises when you are about to undergo bypass surgery and the doctor informs you that a trained healer is assigned to pray for you.

"One reaction is, 'Oh my goodness, they're praying for me, I must be really sick. Why am I chosen? This must be bad news,'" Harrington hypothesized. "Or, 'Oh my goodness, I'm now in a study to demonstrate whether prayer is real. I better get well really fast or I'm going to let God down.' Who knows what they were thinking! But people believe there was some kind of psychological backlash within the patients that led to this unexpected result."

Violating the Laws of the Universe

Richard Sloan, a professor of behavioral medicine at Columbia University Medical Center, holds a less charitable view of these prayer studies. I asked him his assessment: Does the prayer of one person benefit another?

"The answer is pretty unequivocally *no*," Sloan said crisply. "There's recent evidence as well as older evidence which suggests there's no impact of distant, intercessory prayer."

In his book *Blind Faith,* Sloan has amassed an arsenal of reasons why these studies—even if they support prayer—should be discarded.[23] First, they don't take into account the prayers of family and friends who are going to pray for the patient going into bypass surgery no matter what, even if the patient is in the nonprayer "control group" and researchers tell the family not to pray. A wife will not refrain from praying for her husband just for the good of science—and Sloan argued that those "supplementary prayers" muddy the results.

Second, he said, the researchers fell into the "sharpshooter's fallacy," where you first empty a six-gun into the side of the barn and *then* draw the bull's-eye. In this case, the researchers often asked intercessors to pray and then watched to see what happened—declaring victory when, for example, AIDS patients bathed in prayer visit the hospital less frequently.

"In science you have to specify what variable you expect to be influenced by your treatment *before* you conduct your study," Sloan said. "You don't fish around until you find something afterward."

But Sloan's chief complaint is that prayer studies are "a wild-goose chase that violates everything we know about the universe."

"Physicists tell us there are four forces that we know about in the universe. That's it, just four," he explained. There are strong and weak nuclear forces, gravity, and electromagnetic energy. None of these could explain how a weightless thing like a thought or a prayer could affect a person five feet away, much less thousands of miles away.

"There are *no* plausible mechanisms that account for how somebody's thoughts or prayers can influence the health of another person," Sloan said. "None. We know of *none*."

Proponents of distant healing are not troubled by that argument. For years, no one knew how morphine or aspirin or quinine worked. They just knew it worked. Hand-washing was standard medical practice well before anyone hatched a theory of infectious diseases: surgeons and doctors just knew patients fared better when their doctors lathered up. So what if we don't know how prayers help another person? Eventually, they say, we will.

I confess my bias here. My own family life was peppered with these "unexplainable" healings.[24] We did not chronicle these healings except in memory, but over the past century, thousands of other Christian Scientists have done so in the pages of magazines such as the Christian Science *Journal* or *Sentinel*. Perhaps a skeptic could dismiss *some* of the thousands of unexplainable recoveries from near-fatal accidents or Alzheimer's or cancer or other diseases, complete with doctors' verifications. Maybe he could dismiss *most* of them. But *all* of them?

For me, the most satisfying compromise between the proponents and skeptics of prayer came from John Astin, a research scientist at California Pacific Medical Center. Astin reviewed the early prayer studies and concluded that the seeming effects of prayer were real and not just chance happenings. But he also led a study that found that prayer did not help people with HIV.[25] When I visited him, he told me he was genuinely puzzled about the conflicting results.

Astin's hunch is that one person's thoughts or prayers can influence someone else's body. But they may not have the firepower to alter the course of a disease.

"Let's say you have heart disease," Astin proposed. "You've got a whole array of factors that have come to influence your getting that disease. You have biological and genetic factors, family history, dietary lifestyle factors, psychological factors, depression and stress." All of these influence cardiovascular function.

"And let's say I'm a spiritual healer, and I'm trying to influence the course of that disease in some way, to facilitate healing," he continued. "Well, that's not happening in isolation. It's happening within the context of a whole host of complex factors that are influencing that disease. So it doesn't even make sense to think it could supersede the influence of everything else."

In other words, it is impossible to tease out the prayers for a heart patient from his genetic predilection or the thousands of cheeseburgers he consumed over the years.

As I was wrapping up my research about prayer, I realized that science has embraced a sliver of my childhood faith, a century after Mary Baker Eddy "discovered" Christian Science. Most medical scientists now agree that mood states like depression—which are heavily influenced by your thoughts—predict disease progression in a variety of diseases. Or, as my mom would say, your thinking is your experience. Indeed, nowadays scientists shout it from the rooftops, forgetting that they were until recently the snipers gunning for people like Norman Vincent Peale, or Norman Cousins, or my mother.

But positive thinking doesn't require God, and that is the critical point. Many scientists still deride the core of religious belief. They reject that there could be a force that can infuse prayer with power, call it

God or Higher Power or the Divine Mr. Fixit. They reject this because that force, or mechanism, would have to operate outside of the laws of nature as we currently understand them. This is the Maginot Line that separates two sorts of scientists and two sorts of science. Over and over again, I would witness fierce hand-to-hand combat at this very divide.

Later, I learned of a possible—though not widely accepted—scientific explanation for this force. It is called "quantum entanglement"—what Einstein described as "spooky action at a distance." But I had not arrived at that research yet. And so I tackled another personal question. Back in 1995, when my life appeared perfect from the outside and wretched from the inside, I had hit a breaking point and found in that unhappy moment a new spiritual direction. I had always wondered what triggered that dramatic shift, the kind of turnaround or conversion experience that is so common in spiritual journeys. Was it physical, or spiritual, or both? For that, I had to revisit the most exquisite, and painful, moment of my life.

CHAPTER 4

The Triggers for God

WHEN GOD BREAKS INTO YOUR LIFE, it is as if you are lifted up and plunked down in a new spiritual neighborhood. To your friends, you appear unaltered. You still part your hair on the left and speak with the same slight lisp. But you know, if no one else does, that your thoughts and ambitions and loves—your soul—have moved to a new zip code. You're not in Kansas anymore. I can say this from countless interviews and from personal experience. As I heard story after story, I began to wonder what triggered these spiritual transformations. What are the forces that push a person off the cliff of agnosticism and into the sea of faith?

As I searched, I found the usual suspects. Emotional and physical trauma rank high on the list, as does a brush with death. Next to those are the quieter psychological triggers: a poor relationship with one's parents,[1] or stress,[2] or even low self-esteem.[3]

Yet one antecedent stood out, according to the theologians, sociologists, and psychiatrists I interviewed: brokenness. Brokenness occurs when life—in the form of addiction, cancer, singleness, unemployment, or indefinable misery—defeats you. It happens when you come to the end of yourself, you have exhausted your own resources, your own strength and resilience to cope with the situation at hand. You surrender, and in that release, you find a strange calm. It is the only way that many a stubborn soul finds God.

George Valliant, a Harvard psychiatrist in the twilight of his career, put it this way. He told me he used to believe that spirituality could be explained by Freudian principles, or temperament, or as a response to stressors such as poverty. He followed two groups of men from the time they were eighteen—Harvard graduates and men from inner-city Boston. After chronicling their life journeys for more than six decades, he came to a radical insight about spirituality.

"Looking at those who are spiritual, it has nothing to do with mental health, and nothing to do with good fortune," Valliant said. "It has everything to do with recovery from *brokenness*."

Of course, many people encounter the "divine" without a psychological trauma. Young people in particular seem to embrace God without the usual upheaval, which is why evangelistic groups like Young Life or Campus Crusade for Christ focus on teenagers and college students. But for those who emerged from youth untouched by the spiritual, I believe brokenness lies at the root of conversion.

The trouble with studying spiritual experiences is that they're sly little devils. You never know when one is going to sneak up on you. You can't schedule one for next Tuesday afternoon at two-thirty and then keep a log of all the emotional and neurological blips that occur in the days before your encounter with God. This research is by necessity an anecdotal and not a statistical affair.

Therefore, I was limited to post facto anecdotes. The question was, where do you locate those stories? Well, if you want to find bees, search for a honeycomb. If you want to find a cluster of ordinary Americans who have enjoyed dramatic, sometimes physical encounters with God, visit your local church or synagogue. But walk past the minister or rabbi studying in his office. Ignore the faithful singing in choir practice. Make your way down to the basement at seven-thirty on Wednesday night. This is where the transformed people are, the people who have been broken and repaired through spiritual experience. This is where the addicts meet.

The Sudden Alignment

Just as I arrived at Alicia's rambling ranch house on the outskirts of Albuquerque, a minivan pulled into the driveway. Before the wheels had fully stopped, two curly-haired blond boys, who looked to be ten and twelve, leaped out. A trim man in baseball hat, black T-shirt, and jeans emerged from the driver's side. But I was watching the woman who stepped out from the passenger's side, who waved at me and strode over with purpose. For the domestic normalcy I had just witnessed was built on an entirely shattered spirit.

Alicia grew up in a safe, middle-class world, in an intact family that lived in the same house for her entire childhood. There were no obvious dysfunctions, until Alicia became an alcoholic. She was nine years old.

"I remember after everybody went to bed one night," she recalled. "I went into the living room to look at the glasses and the crystal, and it was all so pretty. And I picked out this bottle that had blackberries on it, and I drank it. And I drank it until I passed out. I drank alcoholically from the very first time. But I also remember that feeling of release, that warmth, and that complete feeling of being okay for the first time."

Alicia has little memory of the ensuing years, a testament to the analgesic quality of alcohol, until she dropped out of school in ninth grade. At fifteen, Alicia moved out of her family's house to an apartment, sharing the $235 rent with other high school dropouts "who wanted to drink like I did."

Soon, however, Alicia was outdrinking her friends, and one by one, they dropped away. Then she met Luke (not his real name), another teenager who was able to keep up, and their mutual addiction bound them together. Eventually they married, and the next decade floated by on a sea of scotch.

"It was just one big long drunk," she said. "Round the clock. I didn't have any days of sobriety at all until I was pregnant with my older son. I was twenty-six. That was the first time I was ever clean."

I gazed at this woman, with two boys and an adoring husband, with her own real estate business and her razor-sharp mind, sitting docilely at her kitchen table. It was like watching Pollyanna utter a string of obscenities, so at odds was Alicia's past to her present.

Staying clean during her pregnancy fueled the illusion that she was not an addict; she could stop at will, after all. Luke continued his affair with cocaine, however. A month after she gave birth to their son, Alicia began using cocaine and painkillers, careful to stay away from her true nemesis, alcohol. During this period, Alicia ran a women's clothing store that routinely ranked in the top ten of the four-hundred-store chain. Her external life revealed no fissures. Then she became pregnant with her second son.

"And that was where I started to get to my bottom," she said, "because in that pregnancy I couldn't stay clean. I had crossed the line. And I knew I was really sick."

The spiral downward continued after her second child was born, as Alicia nursed her infant and tried to care for her toddler. She could barely breathe for the chaos in her life and in her head. She and Luke would spend most of their take-home pay on drugs, calibrating their highs and lows with a mix of uppers, cocaine, and alcohol. She had lost her job, she was stuck at home with crying babies, and she began to plot her suicide.

"And I felt completely . . . completely *broken*. I got angry. How could this happen? That's when I said, '*Where are my angels?*' "

In retrospect, Alicia draws a straight line from that desperate moment to a sublime one that occurred a few days later.

One Friday night in May 1996, Alicia and Luke were buying groceries at Costco.

"It was payday, and he had already spent all the money on drugs. We had loaded the bag and we were in line and he told me, 'You know, we can't pay for these.' And I looked at him. And we had to leave the whole basket of groceries there. We came home and I remember the dishes were piled up in the sink. I just remember laying my head on the side of the sink and feeling the coldness of the sink right on my forehead."

She paused, envisioning the moment.

"And then all of a sudden something literally went through my back and my inside. This *alignment* took place inside. And it started down low, like in my stomach and in the lower back, and it was just like my spine was being straightened out. It's like when a cat gets scruffed by its mama on the back of the neck and they get kind of lifted up. And all of a sudden, I knew I was just done. That was it. I took the kids to my mom's, and came back, and told Luke I was going to get clean and sober, and he had to go. And I was in rehab a couple of weeks later."

"What would you say it was?" I asked. "I mean, was it a force from the outside? Was it God?"

"It was an energy," Alicia replied. "I guess people who call that kind of energy 'God' would say it was God. I call it 'soul.' I think my soul got righted at that moment. And everything changed."

Alicia has been clean of all addictive substances since that moment a decade earlier. Luke returned, sober, nine months later, and has remained clean as well. She never went to college but has not suffered for the lack. The real estate business she started made a profit the first year. Her rambunctious sons became national competitors in karate, their trophies displayed on practically every inch of flat space in the house. Raised a Roman Catholic, Alicia forged her own sort of spirituality for years before settling, most recently, on Sufi mysticism.

The one constant in her changing world is Alcoholics Anonymous. This movement has rescued millions of lost souls from the gutter and mended families that have been torn asunder for years. It is built on admitting one's brokenness, surrendering to a Higher Power, and experiencing a "spiritual awakening." In a society that demands double-blind studies and scientific explanations, AA remains stubbornly mystical. It relies on a "God" of one's own understanding to reach down and help the addict defeat his vicious disease. It is the most spiritual of all recovery programs. It is also the most successful.

Prelude to a Shift

Social scientists have studied AA at the safest and most superficial level—the *communal* part of it—and concluded that having a supportive group to which one is accountable aids in your recovery. The same goes for church and book clubs. Astonishingly, no one had studied the *spiritual* element of AA—that is, until a Ph.D. candidate at the University of New Mexico began posting fliers at AA meetings around Albuquerque. In this way, Alyssa Forcehimes conducted one of the most illuminating studies into spiritual experience, finding consistent themes before, during, and after the spiritual moment occurred.

Forcehimes is a beautiful, petite twenty-something whose groundbreaking research led her to spend untold hours interviewing recovering alcoholics about their spiritual transformations. She has an earnest way about her, and a knack for drawing out personal stories, which no doubt came in handy. I asked Forcehimes whether she noticed common themes before the transformation.

"They knew they were falling apart, and that they could not sustain this way of life anymore," she said. "There was a brokenness. And then there was some sort of resolution, like, I have to *do* something. I have to do *something different*."

The spiritual experiences themselves varied, she said.

"They ranged from a 'struck by lightning' experience to dreams that spoke so profoundly to the person that they woke up and changed. Some had an inner dialogue with God, others felt like the weight had been lifted—and they meant that physically, not figuratively."

Seven out of ten people responded to that moment at a physiological level: they felt something change in their bodies. One out of five heard voices or music; and one out of seven had visions or saw a light. In these visceral transformations, I heard echoes of my modern-day mystics like Sophy Burnham and Susan Garren, and Bill Miller's subjects in his book on "quantum change." And like those people, Forcehimes's addicts identified their encounter with the supernatural as the pivot point of their lives.

"They saw the world in a new way," Forcehimes recalled. "Colors appeared different. The world appeared brighter. People appeared friendlier. Many of these people were on the brink of suicide. Prior to the experience, they really did not think that they would be around much longer. They were looking for a way out, and death seemed a possible solution. But after the experience, they were saying, 'Gosh, this is what I'm here for, this is what I can do.' So the experience was really powerful in that way."

And they never used drugs or alcohol again.

Most Americans do not find themselves in Alicia's position—an alcoholic mother with a cocaine-addicted husband and no money to feed her two small children. Most people are not bankrupt or suicidal or disabled from a terrible disease or accident. And yet, many people claim they have been bowled over by something they consider supernatural. For these people, the trauma that leads them to an encounter with God is softer—an aimlessness, or unexplainable hopelessness—the kind of despair that Sophy Burnham felt when she looked at her perfect life and said, *Is this all there is?*

A confession here: I have more than a clinical interest in understanding the prelude to dramatic spiritual experience. I want to know what happened to me.

And My Heart Was Strangely Warmed

I told you that I was assigned to write an article for the *Los Angeles Times Magazine* in June 1995. I did not tell you that at the time, my inner life was a sickening storm of misery. First, my career—always a bedrock of security and self-worth—tottered uncertainly. A year earlier, I had taken a leave from my eleven-year reporting career at *The Christian Science Monitor* for a fellowship at Yale Law School, and finally decided to leave the news business for good. I had received a book contract to write about Burma's Nobel laureate Aung San Suu Kyi, and was preparing to move to Rangoon, which almost everyone I knew thought was professional suicide.

My personal life was equally uncertain: thirty-five and single, I had met a marvelous fellow whom I expected to marry. Unfortunately, it was for the wrong reason: I could already smell the loneliness of middle age and was determined to stave it off. At bottom I knew the relationship would end on the shoals, either before or after the wedding vows.

Most disquieting of all was my crumbling faith in the religion of my childhood, Christian Science. Everything I cared about deeply—my relationship with my parents, my friends at church, my job at *The Christian Science Monitor*, the metaphysical worldview that steered my thoughts and actions—all this threatened to topple once I admitted that I no longer had the energy or fortitude to believe in Christian Science.

I felt as if I were in an operating theater, watching as those parts of my life that defined *me* were surgically removed. A snip here, and my career is removed; a slice there, my faith lies in ruins; a third incision severs my hopes for marital bliss. The surgery was complete. I could no longer identify who I was, for all the distinguishing parts had disappeared. Into that vacuum rushed the shrill questions of an untethered soul: What is my purpose in life? Will I ever have a family, or will I end up a moderately successful, tired woman who eats cereal for dinner alone each night? Mainly, I wondered, *Is this all there is?* They were always there, these questions so common as to be comical—and they drained me of all joy like a dull toothache.

This was the backdrop for my trip to Los Angeles for the *Times* article. These were the questions in the back of my mind when I met Kathy Younge at Saddleback Church, when I listened to her story of cancer and hope, when I sensed an unseen but palpable force as we sat on a bench in the dark, cool night. I returned to my hotel and the next morning bought a Bible. I wanted to know the source of Kathy's serenity in the midst of cancer, and so I began to read the biography of Jesus, beginning with the book of Matthew.

There is a reason the Torah and the New Testament and the Koran are deemed sacred. When they are read at the right time, they can exert a seemingly physical power. The Gospels—which tell of the kindness and boldness and humanness of Jesus—reached up and grabbed me, demanded that I pay attention. The words of Matthew the tax collector, Mark the itinerant, Luke the doctor, and John the fisherman hijacked my senses. I *heard* the voice of Jesus saying to the prostitute, "Neither do I condemn you, go and sin no more"—I heard it as if he had uttered those words to me. I *tasted* the salty air of the Galilean Sea and *smelled* the fear of the fishermen caught in a vicious squall. When Jesus touched the desperate leper, I recoiled from the brackish wounds. This two-thousand-year-old story sprung, like those pop-up birthday cards, from two dimensions to three—from myth to concrete reality.

What unnerved me was that this feeling seemed to come from outside me, not within: it was as if someone had tied a rope around my waist and pulled me, slowly and with infinite determination, toward a door that was ajar. Over those next few days in Los Angeles, I grew curious—inordinately curious—about how these Christians I interviewed each day came to "know" God. What was the password, the *open sesame* that unlocked the mystical door to God? I determined to find out, and as I interviewed people for my *Times* article, I also collected "testimonies"—the stories of people's conversions—hoping to find the combination to the lock. My curiosity became an urgent thirst. I had sipped something mystical on that chilly Saturday night with Kathy Younge, and I wanted more. It was like a long day at the ocean, where nothing matters—*nothing*—except feeling long, cold drafts of water glide down your throat. King David captured it nicely: *As the deer pants for the water brooks, so my soul pants for Thee, O Lord.*

On June 14, 1995, around two o'clock in the afternoon, I lowered my guard. I opened myself up just barely to the notion that there might be a God who cares about me in the same way that Jesus cared about, say, his friend Mary. I prayed—and in that split second of surrender, I felt my heart stir and grow warm, as if it were changing. It was a physical thing, exquisite, undeniable.

Years later, I would learn about a far more famous heartwarming. In 1738, the British Anglican minister John Wesley was flailing about in his ministry, his faith intellectually strong but spiritually comatose. On May 24, he attended a small meeting of charismatic Moravians, who were known for their recklessly joyous faith. At that moment, a door inside Wesley unlocked, emotionally, spiritually. That moment left a visceral fingerprint. Later, he wrote five words that captured the touch of God: "My heart was strangely warmed."

So it was for me. The moment was seared into my memory, and later, when I wondered if I had really encountered God, that warmed heart acted like a Polaroid snapshot, confirmation that a spiritual transaction had indeed taken place.

I wish I could tell you that I was blinded by a piercing light, as Saul was on the road to Damascus. I wish I could that say I smelled roses, the aroma that mystics inhaled in the presence of God. I wish I could tell you that I heard a roaring in my ears, or words, perhaps, like the few simple, ghostly words that Augustine caught when he opened his heart to God. My encounter with the unseen was not nearly so dramatic—not then, at any rate—and yet that quiet moment whipped me around with hurricane force. It became the continental divide of my life, the line that separates "before" and "after." In the next several weeks, the colors I saw were almost unbearably vivid—the cobalt Los Angeles sky at night, the husky green of the summer grass. Even the tan upholstery of my car was so gentle and calflike, I ached to touch it. Things I once enjoyed became dust in my mouth. I walked out of *The Bridges of Madison County* because I could not bear the pain on Meryl Streep's face—she was an *actor*, for Pete's sake, but I could not tolerate even fictional sadness.

And I was terrified by the implications of my decision. My family jokes that all truth can be found in the movies. Perhaps, perhaps not, but the most authentic spiritual moment I have witnessed is not in church but in the movie *The Apostle*. Robert Duvall plays a Pentecostal preacher who murders his wife's lover and flees to Louisiana. He starts a new church, and his flamboyant, holy-roller sermons on the local radio station draw devout, poor, black parishioners.

No sooner has the dilapidated church been repaired and painted than a redneck crashes the services and threatens the parishioners. He arrives during a church picnic on a bulldozer, prepared to mow the church down. Duvall places the Bible on the ground, in his path, and the irate man leaps off the machine. When he stoops to snatch up the book, the preacher kneels behind him, gently placing his hands on the man's shoulders. Quietly, the man begins to cry.

"I'm embarrassed," he whispers, and then he surrenders to God. The man weeps for all he is leaving behind—his friends, his tough image, the internal compass that guided his life. He weeps in terror for the future—where will he fit into the world now, now that his north is no longer true? He weeps out of utter humiliation, a strapping man lying facedown in the dust before the dark-skinned worshippers he has so long despised. It was the most realistic conversion I have ever seen.

I say that because this is what I felt. I, too, had much to lose by my spiritual transformation that day in Los Angeles: my friends, my ambitions, the educated image I held of myself and presented to others. Would I have to check my intellect at the door and take everything on faith? How was I to survive, a believer in a world of skeptics, particularly *journalists*? How would I navigate my world with this new spiritual compass? And at what cost?

In the next few days, I would get a glimpse of the price of wholesale spiritual upheaval. My beau, Steve, was moving to Burma in two weeks. He had managed to find a job there so that we could be together when I wrote my book. A day after this encounter with God, I telephoned Steve from Los Angeles.

"Steve," I said in a tremulous voice, "I can't go to Burma with you. I found God."

An eternal silence ensued.

"Aw, honey," he said. Another long pause. "I was waiting for this to happen." Pause. "Your timing's off, but I'm happy for you."

Steve would spend the next four years in Burma. He eventually found someone else to love. As for me, it would take me another seven years—past the time I could have children—to find my wonderful husband.

My spiritual transformation blinded me with its intensity, a sunburst eclipsing the stars; it took some time for my eyes to adjust back to the more bearable, and dimmer, hues of daily life and work and love. God outshined all other relationships, which is why Sophy Burnham's words years later gave me such a jolt. "How can I compete with God?" her husband had asked. And she responded, "You can't."

A Conversion of Body and Spirit

Ever since my own encounter with something mystical, I have wondered about the physical nature of these moments. I was relieved to find I was not alone: almost every person I interviewed spoke of the light or warmth or the physical touch at the moment they connected with the divine. Sophy Burnham heard a voice; Arjun Patel saw Buddha's eyes; Alicia felt a shock of warmth straighten her spine.

These stories made me wonder how a scientist might explain these reactions. Perhaps that jolt of warmth that burned through Alicia's body was a brief break with reality. Perhaps it was a mental delusion brought on by years of physical and emotional stress. Or perhaps it was the fingerprint that proves God was there.

I called Patrick McNamara, a neurologist at Boston University's School of Medicine and editor of *Where God and Science Meet*, three volumes on the latest research exploring the science of religious experience.[4] I wanted to understand what was taking place at a biological level to bring Alicia or Sophy or me to—and through—that breaking point.

McNamara explained that no scientist can say with certainty what happens during these moments. No one, as far as we know, has ever hit bottom and undergone a spiritual experience while he happened to be in a brain scanner. Still, we do know a few things about the biology of stress and the neurology of meditation, and this allows scientists to make an educated guess.

Some scientists believe that just below a person's well-tended exterior lies an underbrush as dry as timber, strewn with stress, trauma, or a search for meaning. These psychological states are quietly waiting, primed for a spark to ignite a spiritual blaze. That spark can appear insignificant or even random—a stressful day at work, a Zen haiku, a long-forgotten song—but it is enough to set in motion a chain reaction. As in the more dramatic cases, the prelude to this transformative moment is not just emotional. One's body plays a part as well. Something is happening in the mind and in the body, in the psyche and in the physiology, at an emotional and at a cellular level. And at some point these two states, interacting, bring the person to a tipping point.

"There's a whole series of stress hormones, so when the mind interprets a set of events as negative, the stress hormones get released," McNamara explained. "And they function to recruit all kinds of chemicals to meet a threat. In the short term, these chemicals make you stronger and sharper and more vigilant. But in the long term, when the vigilant state becomes chronic, this leads to tissue damage and prolonged activation of the epinephrine system, which activates all of the parts of the body that are meant to meet a threat—the brain, the muscles—the hair, even. If you stay in vigilant state for too long, you'll collapse sooner or later."

When a person reaches "bottom"—if she is lucky—certain things begin to happen. The body may "up-regulate." On the basis of meditation studies, some scientists speculate that when people "let go," as Alicia did when she rested her head on the cool kitchen sink, that can set off a chain of events. The anxiety dissipates, leading to lower levels of stress hormones such as cortisol. The person feels less pain and fear, her breathing slows, and she experiences a sense of release and joy. This is associated with endorphins (natural opiates that the body produces), which are best known for the rush of good feeling called "runner's high." The sensation of happiness and euphoria is enhanced by an overall elevation of the neurotransmitter serotonin in the brain. (Prozac, for example, works in the serotonin system to raise the bottom of depression.) Emotional release can also lead to a surge in a feel-good chemical called dopamine.[5]

This goes some distance toward explaining the bodily mechanics of a conversion or a transforming moment. But does that mean that the "encounter" is nothing more than Alicia's body readjusting itself to stress? Or did her body reflect the touch of "God," of something spiritual, just as water forms ripples when touched by your finger? The answer to that question is not a matter of science—not yet at least—but of opinion, of what you believe about the nature of reality.

For people like me, this is not an academic question. After all, my own worldview changed dramatically, and it mattered to me whether I was touched by the Divine, or just plain deluding myself. Was that sensation of my heart warming no more spiritual than, say, the cold sweat a diabetic feels when his blood sugar dips?

These questions simmered in my mind for years, and in April 2006, I spotted an opportunity to answer them. While attending a conference on science and spiritual transformation at the University of California at Berkeley, I met two very smart and very sympathetic medical researchers. At the tail end of the conference, I asked to speak with them privately, and told them about my experience eleven years earlier when I had surrendered to God and felt my heart "warm." They asked that I not identify them, because even tipping one's hat to spiritual things can mar a career in science.

"Physiologically, what happened?" I asked them.

"Probably there was an increased blood flow to the heart, and that made it warm," one explained. "There are two predominant systems that operate. There's the parasympathetic system, which calms you down, and there's a sympathetic system, which revs you up. And so what you're describing, probably, is that the parasympathetic system kicked in."

The researcher also theorized that a hormone called oxytocin might have been produced in the brain. Oxytocin is the chemical messenger that is produced when a mother is bonding with her newborn baby. And this hormone affects different organs—including, she speculated, the heart.

"I think she's right on," the other researcher said, nodding. He said that studies on primates, rabbits, and bulls suggest that oxytocin release is related to love, and some studies suggest it brings on an "oceanic," sometimes mystical, feeling.

"I'd put my money on something that oxytocin is doing," he said. He noted that in my own case, I had been strained with tensions over work, my boyfriend, and some problems with my physical health. Then an insight, a prayer, a moment of surrender, and the parasympathetic system took over, soothing and calming the body.

"We often underestimate the effects of just relaxing your muscles," he said. "A good massage and suddenly, *Hey! The rest of my life is good!* And you probably had pretty powerful muscle relaxation."

"But what does that mean?" the first researcher asked. "So there's a physiological correlate of whatever is going on. So what? That still doesn't explain it."

Both researchers nodded, acknowledging a central tension in this line of inquiry: science may explain the *biology* of spiritual experience, but it cannot explain the experience away.

A Journey, Not a State

Neurologist Patrick McNamara warned me against pinpointing trauma as the primary trigger for spiritual experience.

"I agree that stress and distress and pain and suffering can certainly lead to spiritual experiences," he said. "But in my experience, interviewing lots of people, and working with lots of different patient groups and reading the literature, I think the thing that consistently leads to spiritual experience is a concerted effort on behalf of the individual to grow in the spiritual life."

He took a deep breath. "In other words, spirituality or religiosity is not simply a response to stress or brokenness or pain or suffering or joy, or any particular emotion. It can be a goal."

This suggests another intriguing catalyst for spiritual experience, one that differs radically from gnawing stress and sudden pain. It is not a state but a journey. It is the search for meaning.

"The human creature is built to need a purpose in life," observed psychologist Ray Paloutzian at Westmont College in Santa Barbara, California. "And if someone doesn't have one, they're going to invent one. And one of the ways that humans do this is by adopting a religious worldview, or some kind of spiritual approach to the world. If they don't have it already—or if they're not comfortable with the one they have—then, because it's a need, they're more prone to spiritual experience."

You might think that spiritual experience should be in the same category as, say, fantasy football or learning Italian—a hobby to brighten one's otherwise dull routine. Or you could conclude that spirituality is a mild psychosis that one develops to cope with reality. However you view it, one of the most powerful triggers for dramatic spiritual events is the desire for it, the kind of search that keeps one up at night, wondering about the universe.

When I began this book, I instinctively knew that my search would reveal a multifaceted "Other," one perhaps more defensible than a list of attributes I memorized somewhere in my Judeo-Christian upbringing: God is love and spirit, omnipresence, omnipotence, and omniscience, to name only a few. It took no more than a few weeks investigating spirituality to stumble upon three surprising characteristics of this "Other" not found in any catechism: God as master craftsman, God as chemist, and God as electrician. The next leg of this journey, then, leads through the hard, sometimes reductionist, science, which in my opinion does not dismiss the notion of God but informs it.

By "God as master craftsman," I mean the one who organizes a person's genes into an astonishingly intricate code. Those genes give a person blue or hazel eyes, make him shy or extroverted, literal-minded or spiritual. God as craftsman helps me with a puzzle that has gnawed at me for years. Many people suffer traumatic experiences, but only some encounter what they believe is another reality. On the flip side, some people experience another reality without a psychological trigger. Why are some people predisposed toward God and others not?

Enter the "God gene."

CHAPTER 5

Hunting for the God Gene

DON EATON TOLD ME he was born wondering about the universe. But he traced his first conscious memory of this passion back to his twelfth birthday, when he received a telescope.

"It was a little five-inch telescope. And I remember the first time I took it out the back door and actually saw Saturn, with the rings on it," he said. The immensity of space—of God—overwhelmed him. "And even at twelve years old I was thinking, How can you capture *spirit*? This is way too big for words and thoughts, it's just all mystery."

Don was in his fifties when I spoke with him. His face was weathered, and laugh lines framed his wire-rim glasses. His goatee was flecked with gray. Don's smile came easily, quietly, the smile of a man who knows a secret, which he would be glad to tell you if only you asked. And his lineage had a mystical strain. His father served as a minister in the United Church of Christ.

"My father's ministry was more a ministry of asking the question rather than coming up with answers," Don told me. "I'd go to him with a question about spiritual matters and he would say, 'That's too small a question.' Whenever I'd ask him for the answer, he'd say, 'Well, my take on it is this, but you've got to come up with your own faith, because if you believe what other people tell you, you have their faith, not your faith.' "

The spiritual wanderlust, it seemed, was passed down from one generation to another. For Don, the search culminated one afternoon in 1980, after lunch, when he was driving along the freeway near Boardman, Oregon, in his 1976 Datsun. It was ugly landscape, rocks and dirt. Don was an itinerant musician and activist. He had recently completed a three-week tour of Oregon high schools, a thankless job where he had been playing his guitar and lecturing about peace and justice to groups of cynical teenagers. He was thirty-three years old, married, underpaid, and on a perpetual search for spiritual answers. At that sweaty, fretful moment, his future seemed as bleak as the landscape.

"I was hot and tired and bored. And I was going, 'Oh, man, I've got another four-hour drive to get home.' I was just thinking, maybe I should get a job with benefits and a regular paycheck. I was feeling really discouraged. And suddenly this thing just happened."

Don poured these musings into a handheld tape recorder. When he traveled, he made audio letters for his friends, which back in 1980 was cheaper than a long-distance phone call. Gradually, the brown ocean of earth before him began to glow, as if someone had turned up the "brightness" knob on his television set. Then, in an instant, the world exploded into pixels of light. His hands began to shake, his breath became shallow. Don clicked the recorder off and veered to the side of the road. A few moments later, he reached for the tape recorder again. This is his account, and as far as I know, this is the only mystical experience narrated live.

"I guess I've just had the kind of experience that Saul must have had on the road to Damascus. All of a sudden, just out of nowhere, I just got a sweeping experience of the Holy Spirit, I guess. That sounds kind of strange coming from

me, because I don't talk like this very often, but I was just moving along and Whammo!

"I went all goose bumps and all the hairs on my arms and legs just started standing on end and I was just kind of full of electricity. Not exactly a voice or anything like that, but kind of a bright, shining message came through to me that said . . . 'Yes, you do understand me, and here is some more understanding,' which is just . . . I'm starting to cry again . . . which is just an amazingly joyful experience for me."[1]

After the vision faded, Don leapt from the car and, crazily, a man possessed, started hurling rocks at the ground, one on top of another, splitting them open.

"They looked like gold to me," Don told me twenty-five years after the experience that still makes his skin tingle. "Everything, *everything*, was beautiful. There was nothing that wasn't full of light—it was that kind of ecstasy. And later, I was reading the Gospel of Thomas, and I came to the phrase, 'Kneel down, break open the rock, you will find me there.'[2] That's exactly what happened to me. I literally broke open rocks, and they were full of light and beauty and everything was infused with God. I just burst into tears."

Don paused, gathering his composure. Then this guitar player in Oregon echoed the words of mystics through the ages.

"It was the actual experience of unity with everything else. It was the classic drop of water in the ocean. But the thing that occurred to me that was miraculous was, I felt the ocean in the drop, not just the drop in the ocean. I felt I was God-stuff—that I was made up of the same stuff that God was made up of, and the only difference is God knew that, and I didn't."

I heard familiar themes in Don's experience: in the moments before the event, a soft brokenness, or sometimes a dark night of the soul; during the experience, the light and the voice, the joy and the unity with all things around him; and after the light ebbed, a radical shift in how he viewed the world that persists to this day. By now, I had heard so many stories like this that I could almost write the script.

But there is something else, which is why I relate this story. From the very beginning, Don seemed *wired* to search out God, to place the eternal questions above all others. It was an urgent and sensory thing for him, and it was obvious to everyone else—his family, his wife, even the friend to whom he later described his experience.

"And I said, 'Man, I don't know why I was blessed to have this experience, because it's certainly nothing that I did. I was just driving along the road,'" Don recalled. "And my friend just looked at me and said, 'I don't believe that for a second.' I said, 'What do you mean?' And he said, 'You have been knocking on the door one way or another as long as I've known you. You've been on a spiritual pilgrimage, you've been on a deep quest. I think the experience happened to you because you've been priming the pump for years.'"

The question for me is, *why* did Don Eaton embark on a spiritual search from the time he was a boy? Why did he look through the telescope and see spirit, while others see the Big Dipper? Did he experience unusual transcendence because he pursued it—and if so, why did he want to pursue it in the first place?

Then there are the other mystics I interviewed. Sophy Burnham jumped with both feet into Christianity, Hinduism, Buddhism, and meditation, before going to the mountains of Machu Picchu. What drove her? And why did she hear the thunder of the universe, while the other tourists heard a history lecture about the fortress? Why did Arjun Patel see Buddha's eyes when he meditated, while his friend sitting a few inches away did not?

When I asked Arjun this question, he shrugged in genuine puzzlement.

"I don't know why it happened to me," he said. "But I'm pretty convinced it has nothing to do with me being a special person. The only qualification was that I was a human being."

Yes and no. Arjun was pursuing a spiritual life with more verve than the average college freshman. Realistically, not many eighteen-year-old men meditate every day. An internal rudder steered him that way.

You need not visit an ashram in India to recognize that some people are more spiritually inclined than others. The question is, *why* are some people spiritual virtuosos and others spiritual duds? Here the evidence is just emerging, but the research suggests that spirituality is genetically "soft-wired," like intelligence or a gregarious personality.

The search for those religious genes has barely begun. We are a long way from building a neuroscience of religiosity. But the concrete has been laid and you can begin to see the superstructure of this new edifice. I will begin, then, with the foundational work: the notion that spirituality runs in the family.

A Knack for God

A few years ago, shortly after I had joined National Public Radio, my mother went to the Kennedy Center in Washington, D.C., to hear the National Symphony. She was freshening up in the ladies' room when in walked Nina Totenberg, NPR's Supreme Court correspondent and one of the most famous of my colleagues. My mother identified herself and said, "Oh, Ms. Totenberg, I do so admire your work."

The next day, Nina dropped by my desk.

"Barbara," Nina said in her operatic voice, "I met your mother yesterday. She's so . . . *spiritual-looking*."

Nina was right. My mother *is* spiritual. To my mother, events and even physical objects have eternal significance. They are Plato's forms that are only a shadow of the real, unseen world. As a devout Christian Scientist, Mom relies on prayer for everything from finding a lost earring to recovering from the flu. Although I am no longer a Christian Scientist, I seem to have inherited her transcendent view of the world.

And although no one would describe me as "spiritual-looking," I did for a few seconds peer, as she did, at the misty border of another dimension, and emerged, as she did, fundamentally changed. Of course, it is entirely possible that Mom did a very good job of training me to be spiritual—nurture at work. But I have often wondered, what about nature? Could genes be at play?

It makes sense that some of Mom's spirituality landed in my genes, since half my genetic coding comes from her. I decided to try to measure how much our sensibilities overlapped. I called psychiatrist Robert Cloninger at Washington University Medical School in St. Louis. Cloninger has developed what has become the gold standard of personality tests.[3] His paper-and-pencil questionnaire measures, among other things, "self-transcendence," or an inclination toward spirituality. That includes one's belief, or disbelief, in phenomena such as mystical experience, or a belief in miracles, the supernatural, and a force greater than oneself directing one's life.

The four dozen or so spirituality questions are embedded in 240 questions, so you cannot really know which ones are aimed at measuring one's propensity toward God beliefs or, say, optimism. After my mother and I took the test, separately, we found that our answers about spirituality were identical. I was impressed. Of course, that does not prove anything. But it is suggestive.

Certainly, I wondered whether this was a rigorous enough instrument. And one night at dinner, my husband—a political scientist and an expert on South Asia—asked the obvious.

"Is anyone concerned that self-reporting questionnaires are notoriously unreliable?" he asked. "If I wrote an article about India and said, 'I interviewed the Indian government and they said they were into peace and disarmament, that they had no desire to threaten Pakistan with nuclear weapons'—if that was the extent of my research, people would question its validity."

Scientists admit this is the chink in the armor. Indeed, the test itself implied this weakness. Question 230 is: "I have lied a lot on this questionnaire. True or false." But at this point, it is their best methodology. The self-transcendence questionnaire has become, for now at least, the basis for not only statistical twin research but genetic analysis as well.

A few days later, I told Nathan Gillespie about the identical results my mother and I had scored on the spirituality test. Gillespie is a researcher at the Virginia Institute for Psychiatric and Behavioral Genetics in Richmond, and he has been thinking about the interplay between genes and spirituality for some years now.

"Spiritual beliefs tend to clump in families," he replied. "That's one of the questions we're interested in: What makes some individuals more spiritual than others? Why does spirituality aggregate in families? Why does the apple fall not very far from the tree?"

To unravel that mystery, scientists look to twins—in particular, identical twins, who come from the same embryo and share virtually all their genetic makeup. (Recent studies suggest that there are tiny variations in the DNA of identical twins, but they are minuscule.)

Of course, social environment or upbringing clearly plays a part in one's spiritual yearnings. Moreover, researchers believe that *which* religion a person practices has almost nothing to do with DNA, and everything to do with parenting and culture. A person joins a Southern Baptist church not because there's a Baptist gene but because he grew up in Mississippi going to his parents' Baptist church. The same is true of the Hindu girl in New Delhi and the Shia Muslim boy in Tehran.[4] Still, how *intensely* a person believes in Jesus or Vishnu or Allah is guided at least in part by his or her genes.

Let me pause here, because the words that I have so blithely written represent a tectonic shift for me, and perhaps for anyone raised in a particular religion. While religions make claims to doctrinal truth, genetics points to the capacity in each of us to experience transcendence, to envision and connect with another dimension. Genetics—and science in general—cannot referee between Christianity and Islam, or

Buddhism and Zoroastrianism. It cannot determine winners and losers. It can explain the mechanics of satellite transmission but it cannot say whether ABC or NBC has better content.

This is not to say that the scales fell from my eyes and I suddenly saw that all religions are one and the differences can be handled in footnotes. Far from it. Rather, I realized that I needed humility as I thought about what faith I embraced. This was dangerous territory for me as a Christian, but I could not shake off the feeling that there might be many vectors to Truth.

The Case of Identical Twins

Let's say there are two families, the Joneses and the Smiths. The Jones boys are identical twins; they share virtually 100 percent of their DNA. The Smith girls are just sisters; they share 50 percent of their genes. If DNA had *nothing* to do with the children's spirituality, their religious intensity would be determined only by their upbringing and the events that happened to them—a run-in with a particularly vicious nun in Catholic school (which turns one person against faith), or an inspiring Pentecostal service (which turns another on to it).

If, on the other hand, genes *do* play a role, then you would expect to see the spiritual inclinations of the Jones boys matching each other more closely than do those of the Smith girls, because the Jones boys share twice as much of their genetic makeup. It doesn't work to test this on an individual scale, but when scientists persuaded hundreds or thousands of twins to participate in a study, a clear trend did emerge. In study after study, researchers found that between 37 and 50 percent of the variation in spirituality appears to be explained by genes.[5]

To tease out the contribution of upbringing and life events (nurture) from one's genetic wiring (nature), scientists have looked at a tiny subset of identical twins: those separated at birth. For those twins, environment (or nurture) cannot influence the siblings to be more like each other, since they were raised in different families and did not share the same environment.

Unfortunately for me, twin researchers didn't include a category in their database for "identical twins separated at birth with strikingly similar spiritual intensity." But fortunately for me, Nancy Segal, who heads the Twins Studies Center at California State University, Fullerton, happened to know Debbie Mehlman and Sharon Poset.

In 1952, when they were one week old, the twin sisters left a New Jersey hospital with different adopted families. For more than four decades, neither one knew that a genetic carbon copy existed somewhere in the world.

Debbie's family moved to Connecticut, where she married and lives today. Sharon eventually moved to Kentucky. Only when they were forty-five years old did Debbie's adoptive mother reveal that she had an identical twin.

In retrospect, it made sense, Debbie told me. "We both always felt there was something missing."

Debbie hired a private investigator, and two months later, she and her sister, Sharon, met at the airport in Hartford. Coincidentally, the two had selected black skirts and beige tops for their meeting. Their accents fell into an identical New Jersey twang. Each had an odd habit of crossing her eyes to suggest irony.

"We did so many things alike, people were clutching their chests," Sharon recalled.

As they pieced their stories together, they realized they shared a number of twists in life: from similar haircuts throughout their teens (not so interesting), to majoring in sociology (a little more intriguing), to raising one child each, both of whom joined AmeriCorps. But it was their spiritual journey that drew my attention. Debbie was Jewish, Sharon was an evangelical Christian, and both had always been drawn to God.

Sharon grew up Catholic; her parents and her other (nonbiological) sister displayed little enthusiasm for the faith.

"The sister I grew up with would go to church and wait in the parking lot and grab a bulletin so my parents would think she had gone to church," Sharon recalled. "But I always wanted to go to church. I got my Holy Communion. I was fascinated with it."

Eventually, Sharon became a born-again Christian, and she now works at an 8,000-member "megachurch." Like Debbie—who teaches in her synagogue and is learning Hebrew—her faith is the center and circumference of her life.

Debbie, too, wondered about her early religious faith.

"I have a mother who's like, 'Oh, I had a headache once during Yom Kippur so I never fast.' And that drove me nuts as a kid!" Debbie laughed.

As in Sharon's case, Debbie was mystified that her nonbiological sister showed little interest in synagogue. "My [nonbiological] sister and I grew up in the same house, and I used to think, How come it means something to me and it doesn't mean anything to her?"

The answer, she believes, lies in their genes. "When we went to school in the sixties, everything was nurture," Debbie said. "It was all 'environment, environment, environment,' and I used to figure, Oh, my mother must just be weird, I must be like my dad, and try to rationalize. And then Sharon and I met, and it's like, Well, there goes environment out the window."

I put the question to Nancy Segal: Is it *all* genes?

"Nothing is all genes," Segal said, "but I think it's genes to a large degree. The content is fashioned largely by the culture, but *how* they're doing things, *how* they're going about living their lives, is very similar. And that has to be something genetically influenced, because they were raised in different places."

Once the twin researchers believed they had established a link between genes and spirituality, it was inevitable that geneticists would try to find which genes made for a Sigmund Freud and which ones led to

Saint John of the Cross. This line of inquiry is in its infancy, with relatively few studies, in part because scientists find it easier to win grants to study schizophrenia than spirituality. When the spirituality studies are conducted, it is usually on the sly, with data gleaned from existing smoking or alcohol-abuse studies. This is how Dean Hamer claimed he stumbled—oh so controversially—upon the "God gene."

The God Gene at NIH

If you want to conduct a spirituality test on very high-end subjects, look no further than Bethesda, Maryland. Dean Hamer and Francis Collins are both leading geneticists at the National Institutes of Health. They both love the scientific method. Where they part company is God; specifically, whether one's inclination toward God boils down to genetic coding or something more.

"Genetics has become so much more powerful," Dean Hamer told me. Hamer is a Harvard Ph.D. who works at NIH's National Cancer Institute. "Things that seemed utterly mysterious, like the pattern of dots on the wings of a butterfly, have now been reduced to individual genes acting at specific times during development. So the idea that something as complicated as 'why people pray' might be studied, has gone from a complete fantasy to something that scientists can think about in a serious way."

Hamer is author of *The God Gene*,[6] a book that has many scientists pulling out their hair in frustration, since they felt the study was shallow, sensational, and published without scrutiny by other scientists. I wasn't sure what to expect when I arrived at his four-story brownstone near Logan Circle, a once scary, now fabulously expensive neighborhood in Washington, D.C. Hamer opened his door, shushing aside two barking dogs. He wore a black T-shirt, tan shorts, and Birkenstock sandals. He looked like a professional soccer player, trim and compact, with salt-

and-pepper hair, neat features, and cocky eyes. He was a man that any woman, and some men, would notice at a cocktail party.

We walked up the spiral staircase to the third floor. A framed copy of the *Time* magazine cover featuring *The God Gene* hung on the wall. A dozen copies of his book were scattered on tables and chairs. We gravitated toward a tiny room of rich grays—gray walls, gray chairs, a small round glass table.

Hamer was a practiced interviewee—engaging, modest, with the charming verbal patter of a springtime shower. He has written that he is a "materialist," which means he believes that much of people's behavior can be reduced to physics and chemistry. What, then, drew him to look for God in the genes?

"When I was a young adolescent, I thought for a while that I might want to be a priest or a minister," he laughed. "But that's because I was such a troubled child and had such a huge case of attention deficit disorder that I just wanted some way to get out of all the trouble I was in at the time. So for me it was probably not so much a deep sense of spirituality as a deep curiosity, combined with a willingness to do stuff that you're not really supposed to do in science. That's why I studied the gay gene, that's why I studied personality. These are things that normal scientists are afraid to do, and I'm just, like, Heck, why not?"[7]

"So you're still that troublemaking adolescent," I observed.

"I'm still that troublemaking adolescent. But now I do it with a DNA extraction kit and scientific tools."

Hamer described his study, which he conducted on his own time, using data collected for his work as a cancer specialist at the National Institutes of Health. He recruited about a thousand people, whom he asked to give a DNA sample and fill out the Cloninger "self-transcendence" test, the same one I had sprung on my mother.

Once Hamer knew which people scored high on spirituality and which scored high on cool rationalism, he looked for variations in their DNA.

Hamer found that the most spiritual people had one small difference in their DNA that was not present in the less spiritual people. The variation was located in a gene called VMAT2, which regulates dopamine and serotonin, both chemicals that affect the way people perceive the world and the way they feel about it.

"Everybody has the so-called God gene, the VMAT2 gene," Hamer said. "But they have very slightly different versions of it. So it's like different flavors of ice cream, some people like chocolate, some people like vanilla. Some people have the more spiritual version of the VMAT2 gene, others have the less spiritual version."

Did this gene alone account for all of the difference in spirituality between two people? No. Half of it? Wrong again. The discovery, it turns out, was somewhat more modest: VMAT2 accounted for less than one percent of the difference in spirituality between two people. It raises the temperature from, say, 45 to 46 degrees, not from freezing to the boiling point. In other words, this is not the DNA equivalent of the Almighty who fills heaven and earth. It's more like Pan, or Persephone, or the Furies: one small gene in a pantheon of genes.

"Clearly this is not *the* gene that makes people spiritual," Hamer conceded, adding that he regretted the title of his book (although that title has surely accounted for robust sales). "There probably is no single gene. It's one of many different genes and factors that are involved. As a biochemist, I don't expect to solve the mystery of spirituality with one genetic analysis. But I do hope that it will inspire other scientists to get involved in this area."

I asked him if his research had influenced his view of God.

"When I started out, I was a typical scientist, completely skeptical of anything religious—you know, 'the opiate of the masses.' As I was researching the book, I made it a habit to do various spiritual or religious things. I went to religious services every Sunday. I spent a week in a Buddhist retreat in Japan, studying Zen. And the more I did this, the more I realized that, yeah, this really is a very powerful thing.

"But it hasn't affected whether I think there's a 'God' or not," he continued. "I don't see any compelling evidence that there is. I have no disproof, either, so I'm your classic agnostic."

This rooting around for God in the genes reminded me of an acquaintance of mine, a neurosurgeon who loved carving tumors out of people's brains. The surgery is so *interesting*, he would say to me, but he never mentioned whether the patient lived or died unless I asked him. As I talked to Hamer, I recognized the same sentiment: whether or not God exists didn't really matter to him; he had no dog in that fight. I wanted to hear from a scientist who did.

More Than Molecules

Francis Collins and I sat in his spacious, sun-drenched office at NIH overlooking a canopy of trees. Collins headed the federal government's Human Genome Project and wrote *The Language of God.*[8] He was a towering man with a tame mop of gray hair and large features that fell automatically into a smile. On this blistering-hot day he wore black jeans and a blue T-shirt with an American flag spanning his chest. It was the Friday afternoon before the July Fourth weekend in 2006. The employees had the afternoon off; most of his colleagues had left, but Collins talked with me for more than an hour.

"There is no gene for spirituality," he told me. "There may be a lot of genes that play some role in the development of a personality" inclined toward God, he said, but "there are so many other things at work."

Dean Hamer's book evidently vexed him. Collins ticked off the flaws. Hamer had declined to publish the findings in a peer-reviewed journal, opting instead to go straight to the public with a popular book. That meant his research did not endure rigorous scientific review and challenge. Collins continued: The findings have not been replicated, meaning that the "spiritual" properties of the VMAT2 could be a statistical fluke. The entire project rested on self-reported questionnaires, he noted. And finally, he said, the "God gene" seemed to account for a tiny difference in self-transcendence.

"Maybe the right title for the book should have been: *The Identification of a Gene Variant Which, While Not Yet Subjected to a Replication Study, May Contribute About One Percent or Less of a Parameter Called Self-Transcendence on a Personality Test.*" Collins reflected on that a moment. "That probably wouldn't sell many books, though."

But isn't Hamer on the right path? I asked. Genes have already been identified with diseases like Parkinson's and Alzheimer's. Why not a gene—or a cluster of genes—to explain spirituality?

"Spirituality is a very different phenomenon. We're not talking about pathology here. We're talking about a transcendent experience," he argued. "You could no more identify a genetic glitch involved in spirituality than you could identify a genetic glitch in the experience of being alive. It is much too complicated and interwoven with every aspect of your personality, of your consciousness, of your sense of who you are, to be able to be pinned down in that very deterministic way."

I knew that Collins was an evangelical Christian, a statistical fluke in the rarefied scientific atmosphere he traveled in, which must surely affect his approach to the research. I was unsure how to broach this intimate question. He did it for me.

"Take myself," he said, out of the blue. "I have been a blatant materialist during a significant part of my life, an atheist with no use for God. I am now a very serious believer. My DNA did not change during that time interval. So I can certainly cite from personal experience that it's nothing about my heredity that changed my view from one end of the spectrum to the other."

For Collins, the path to God wound not through the shroud of mystery but the light of logic. When he was a twenty-seven-year-old medical student at the University of North Carolina, he said, a woman with terminal cancer asked him whether he believed in God and eternity. He evaded the answer, but the question haunted him.

"I came to the realization that my atheism had been chosen without considering the evidence for or against faith," he said. "As a physician/scientist, you're not supposed to make decisions without looking at the data. So I decided I better learn more about this so I could defend my position—and accidentally converted myself."

For a "tortured" year, Collins wrestled with God. He was mortified at the prospect of accepting God. Would he have to become a missionary? Would he turn into a dreary, humorless person? He eventually concluded, reluctantly, that God made logical sense to him: Where else would the moral laws that are "written in our hearts" come from? When Collins finally surrendered, it was like a jet breaking through the sound barrier: the turbulence ended.

Ever since, Collins has let his religion inform his science, and his science inform his faith.[9] He and his team at NIH have mapped the human genome, identifying more than 30,000 genes in the human body. Genetics, Collins said, has not explained what he sees as the central element that makes humans different from animals: the sense of right and wrong, which often prompts people to sacrifice themselves, not just for friends and family, but also for enemies. It is a drive that, according to Darwin's theory, would make one's selfish genes very unhappy. Nor, Collins said, can genes explain the transcendent moment.

Transcendence is like music, he suggested, "where you are transported briefly into this experience that you can't put into words, which leaves you with this longing, a longing to be part of something but you don't know what it is. For the atheist, well, that was just your amygdala [in your brain] going off, I guess. But for the believer, it's a signpost calling you to examine yourself and ask: What's here more than molecules?"

How to reconcile the ideas of Dean Hamer and Francis Collins? Why does one man begin contemplating ministry and end up agnostic, while the other tracks exactly the opposite course? Why does inquiry lead Hamer down to molecules and Collins up away from them? Why does Hamer's view of the science seem intriguing, while Collins's view of the world seems true?

And yet, for all their verbal sparring, I do not think their views of the science are irreconcilable, since both believe that biology must be involved with spiritual feelings. They agree there is a gravel road beneath their feet; the dispute is over where the road leads. Is it a large loop circling back to the same physical spot, or is it a path to a different, spiritual, state? For a moment, let us to stoop down and examine the stones in the road, without worrying about the destination.

Diving for DNA

As scientists began to hunt for genes or brain chemicals that might exist in "spiritual" people, one of the chemicals they quickly trained their sights on was the neurotransmitter dopamine. Dopamine is the "feel-good" chemical in the brain. It is what makes runners euphoric after a long jog. It is what is so addictive about cocaine and amphetamines.

As neurologist Patrick McNamara at Boston University explained it, a gene will "code for," or regulate, certain chemical processes—such as the production of dopamine, which then stimulates different parts of the brain. Think of a gene as a dimmer switch for a light: it can turn it on and off, but it can also give some people more or less of a certain trait. McNamara and others say that if there are "spiritual" genes that code for dopamine, then that dopamine would stimulate parts of the brain that could create a spiritual experience, transcendence, or a feeling of God.

"There are genes that help regulate the levels of dopamine in specific areas of the brain: the limbic and prefrontal lobes," he explained. "And those areas of the brain in turn support all kinds of complex functions. And many of those complex functions in turn would support these more basic capacities related to religiosity, namely positing supernatural agents and engaging in rituals."

The limbic system of the brain is involved in emotions, such as awe, joy, ecstasy, transcendence, deep sadness—emotions that seem to pour out of mystics. The prefrontal lobes involve more complex thought, reflection, and attention—and researchers have found that these areas play a big role in prayer and meditation.

Therefore, theoretically, a gene that codes for the activation of dopamine could affect spiritual feelings and religious behavior. The question is: Is there any evidence of a link between spirituality and dopamine?

David Comings and his team of genetics researchers at the City of Hope Medical Center in Duarte, California, had a hunch that there might be. He speculated that a particular dopamine docking station, or receptor gene, called the DRD4, might have something to do with spirituality.[10] The researchers recruited 200 men in California. Some of them were college students, while others were recovering addicts in a nearby treatment program.[11] They asked the men to complete Cloninger's self-transcendence test, and to donate a bit of their DNA. The geneticists knew that dopamine receptors varied from person to person. Comings and his colleagues hypothesized that the gene variation (or "polymorphism") might affect whether a person believes in God.

That is precisely what they found (although, remember, this is a single study). People with a particular variation of the DRD4 gene scored higher on the self-transcendence scale. To a layperson, that particular gene might seem only modestly important: 3.9 percent of the difference in the men's spirituality scores could be traced back to that particular receptor.[12] However, Comings noted that it is rare for a single gene to account for more than 1.5 percent of variance of any behavioral trait.[13] He also noted that the dopamine receptor is present in high concentrations in the frontal lobes of the brain, which is the site of many higher human brain functions.

"One could argue," he and his colleagues wrote, "that spirituality is the quintessence of higher human brain functions."[14] While the scientists cautioned that this particular gene "is not 'the gene for spirituality,'" it does seem to contribute to a "significant portion" of the variance, or reason, some people are spiritual and others are not.

There is another frequently mentioned suspect in the God-gene lineup: the serotonin system. The serotonin system has intrigued scientists for years because it dramatically alters moods. For example, the drug Ecstasy creates a high by releasing a wave of serotonin. Prozac, Paxil, and Zoloft work more slowly in the system, evening out moods by doing the same thing. And hallucinatory drugs can create a mystical experience worthy of Saint Teresa of Ávila.

In 2003, a group of Swedish scientists led by Jacqueline Borg at the Karolinska Institute in Stockholm tried another approach to determine what role, if any, serotonin plays in spiritual experiences. They used brain scanners. They recruited fifteen healthy Swedish men for a spirituality test. The subjects took Cloninger's self-transcendence test and then sat for a PET (positron emission topography) scan, which took pictures of their brain activity.[15] Researchers cannot measure the amount of serotonin in the brain directly, so they used an indirect method: they measured the activity of the chemical's docking stations or receptors, and in a particular receptor gene called serotonin 5-HT1A.

To do that, the scientists injected a tracer fluid that acts very much like serotonin into each man's bloodstream. Then they put the men separately into the brain scanner and watched what the fluid did once it arrived in the brain. They were particularly keen on seeing whether the counterfeit serotonin would bind with serotonin receptors. Their hunch was confirmed: they found a strong relationship between each subject's serotonin levels and his spirituality score.[16] Specifically, this genetic dimmer switch seemed to affect which man believed in God, a unifying force, or phenomena that can't be explained by "objective demonstration," and which tended to favor a "reductionist and empirical worldview."

This suggests, the researchers wrote, that "the serotonin system may serve as a biological basis for spiritual experiences," and that the variation in this gene "may explain why people vary greatly in spiritual zeal."[17]

Researchers in this area remind me of Sherlock Holmes, piecing together scenarios with shards of often conflicting evidence. It is evident that *something* has happened, but exactly how it happened remains a mystery. So it is with scientists exploring spirituality: they know that millions of people genuinely experience transcendence—but what, exactly, is the mechanics of that feeling? Is it genes, or temporal lobes, or a psychological coping mechanism? Or is there a Higher Being, something that few scientists have considered yet, like the dog who didn't bark?

Does God Play Favorites?

As for scientists like Boston University's Pat McNamara, he's thrilled the chase is on.

"There's going to be a specific biology involved in religiosity," he predicted expansively. "There's going to be specific brain systems that are always involved in religiosity. There's going to be specific neurochemistry. There's going to be drugs that selectively influence religiosity. There's going to be a specific cognitive neuroscience of religiosity. And on and on and on! But that doesn't mean that religiousness is nothing but the biology of religious experience. It just says that religiousness is one of the innate capacities of human beings, like a host of other traits."

"So maybe there's a Coder, so to speak?" I asked. "I mean, if there's a genetic code, is it possible there is a Coder, or a Higher Power, or a Creator?"

"Theoretically, if God wanted to communicate with us, then He, She, or It would create a biology that allows us to communicate," McNamara stated. "So it makes sense that there is specific biology that allows for that sort of relationship. But the fact that there *is* a specific biology of religiosity does not rule in—or rule out—God."

While the findings in this realm are preliminary, they do suggest that some people have a greater propensity to respond to God than others. But, I sputtered, feeling acutely my own spiritual mediocrity, what if there is a God who writes His language in our genes unequally—apportioning to some a bounty of spiritual gifting and to others a meager gruel? Isn't that a bit of deistic favoritism?

"For some people that could be troubling," observed Ron Cole-Turner, a bioethicist at Pittsburgh Theological Seminary. "It's not for me, but it could be for some. If you thought God is going to send some people to hell on the basis of a lack of response and the lack of response is hardwired in our genes—that would be troubling.

"But," he added quickly, "don't read 'genetic' as deterministic. Because genes don't create hardwired, deterministic outcomes. They create a range of possibilities that might incline you one way or another. It's by no means determined that if you have a certain gene sequence you are spiritual and if you don't have a certain gene sequence you can't be spiritual."

In other words, nature (genes and biology) plays some role, but so does nurture (one's environment and life events). What genes do is create a sort of tipping point for spiritual experience. If you are genetically inclined toward spirituality, a relatively small event can flip your Sunday mornings from sitting on the couch watching the Sunday talk shows to sitting in a pew watching a preacher. But if you are born with a genetic predisposition to think religion is complete bunk, then it's going to take an enormous dose of environment to push you to religion.

It's a little bit like automatic air-conditioning. For some people, a relatively modest rise in temperature—breaking up with a boyfriend, for example—can flip on the cooler system. Those people are genetically inclined to be spiritual. Others may sweat it out to 90, 95, 100 degrees; only then will their God-switch flip on. And some would rather die of heat than turn to "God."

Those explanations worked well for psychiatrist Robert Cloninger for many years. Cloninger, who developed the self-transcendence scale, heartily believes that genetics is at play. He has conducted some of the studies himself. He also believes that life events push a person toward or away from spirituality. But the more people he worked with and the more studies he performed, the more he felt he was missing a piece: the soul.

"I realized there were other differences that we didn't measure with temperament or character or genes, that were really spiritual differences," Cloninger said. "There was a whole dimension here that was basically unexplained. Like how do you explain free will? You can't do that from a materialistic standpoint. How do you explain intuitive creations and sudden insights? You can't do that from algorithmic thinking. How do you explain the gifts of Mozart? I had to move to recognizing that we actually do have a psyche, a soul, and that it does have characteristics, and those characteristics differ from one to another."

"You seem to be saying," I said carefully, "that there's nature, nurture, and then something . . ."

"Mystical?" Cloninger laughed as he finished my sentence. "There is. And it's real. We really do have a soul. And we really can listen to it. And it's good and it's intelligent and it's what makes life worthwhile."

To those inclined toward God, this makes perfect sense, while others may scratch their heads in puzzlement. As for me, I see this as an invitation to unravel the mystery of soul and body and their relationship to each other. It is a laughably ambitious task, and to even begin, we have to drill down further into the science. We must wade through the synapses and lobes and chemicals in the brain, the stuff of neurology that is so intricately tuned that it takes my breath away in wonder.

Let's turn, now, to God as chemist. The brain is a cauldron of chemicals that color how we feel and think. Mania and depression, fear and love, all of our moods trace back to chemical reactions. Why not transcendence as well? And if there is a chemistry to spiritual experience, does that imply the hand of a Chemist? These puzzles led me to the ancient practices of Navajos and to the rebellious days of the 1960s. They zeroed in on a particular neurotransmitter that holds a key to spiritual experience. They took me to a type of spirituality that is both instant and measurable synthetic spirituality.

CHAPTER 6

Isn't God a Trip?

THOSE WHO SAY LIFE IS SHORT have never attended a peyote ceremony.

This thought occurred to me just after midnight on July 23, 2006. Thirty-one of us formed a circle around a fire in an enormous tipi we had erected on top of a mountain near Lukachukai, Arizona. Thunder and lightning ripped through the sky, and heavy rain lapped under the edges of the tent, soaking our cushions and turning the dirt floor into a muddy paste. My companions sat cross-legged on the floor, motionless, gazing at the flames with sleepy eyes dilated by the mescaline from the sacred herb peyote. Three strapping young men with long black hair moved around the circle, pounding on a water drum and singing an urgent chant, which sounded to my untrained ears like, *Doo doo doo doo DOO DOO doo doo doo doo DOO DOO.* I understood nothing, since they sang in the Navajo language, Diné, but I felt the power of the chant like ropes wrapped tight around my body. I could not move, only breathe.

But I wanted to move. I was desperate to move. I had been sitting for three and a half hours on a thin, wet cushion. The only relief came in shifting from cross-legged to a kneeling position, and then for but a moment. Unlike my happy and stationary co-participants, I was not in fact stoned. I was an "observer"—an observer of one of the only forms of drug-induced spirituality sanctioned by U.S. law. Psychedelic drugs such as LSD, psilocybin (mushrooms), and mescaline had been outlawed by the "war on drugs" in 1971, ending most of the emerging research on the states these drugs seemed to uncork. Only peyote and ayahuasca used in Native American religious ceremonies are permitted, which is how I found myself sitting in a tipi, bobbing my head to Navajo chants with a silly grin, wet and sore and as close as I could legally come to observing mystical states created by Schedule I drugs.

Well, almost as close. I could have ingested enough peyote to reach an altered state myself. But I had opted out. Okay. I took a little. The law permitted Native Americans to ingest peyote for religious purposes only, and there appeared to be no loophole for NPR. More important, I thought the peyote might actually *interfere* with my work, since violent vomiting is common for the uninitiated. I then reminded myself that I had a stepdaughter, and tripping on sacred mushrooms might not be the best message to send a twelve-year-old. Whole truth be told, I also worried that the peyote would deliver on its promise and thrust me into an enlightened spiritual state. I fretted that God might be reduced to a chemical, making my own daily commitment to spiritual practices look a bit archaic. All that prayer and study, when I could just swallow a bit of mescaline—a little like using the Pony Express in the age of e-mail.

Once in the tipi, I found that skipping the altered state was, from a culinary point of view, less of a sacrifice than I had imagined. When the peyote man first came around with his coffee can full of dark brown sludge and spooned the peyote paste into my mouth—using the same teaspoon for everyone, I noticed—I nearly choked on the acrid taste and lima-bean–like texture. Just as I was recovering from the paste, another man thrust a silver bowl in front of me. I reached into what felt like a mass of writhing worms and plucked out a peyote button, the cactus herb itself. I held the soggy yellow button reverently in my hand until he had moved on, and then quietly dropped it on the dirt behind me. I looked up to find a third man kneeling before me with a jug of green liquid—peyote tea—which he pressed to my lips before I could brace myself.

The trinity of peyote would return every two hours or so, but after the second circuit I politely declined, leaving myself in a wired but not altered state. At just after midnight, the drumming ceased, catapulting us into silence, save for the hiss of the fire. Eventually the Navajo woman in the place of honor cleared her throat, and we all turned to gaze at her.

"I want to thank you for praying with me," Mary Ann began in a reedy voice. "I know that the peyote and your prayers will heal me. Now I want to tell you something I have never told anyone." She paused, looking around the circle. "I need to confess to the fire."

Jesus as Chemical Compound

I had met Mary Ann the morning before. She showed up just as a group of us—about a dozen Navajos, a Harvard professor, his wife, and I—were heading off to set up the tipi for Mary Ann's healing ceremony the next night. She had come to work out the last-minute details.

My instant impression was that this was a woman in long-term pain: She was sixty-four, and her golden face bore deep lines around her nose and mouth, which was set in grim resignation. She jackknifed slightly at the waist, bent in discomfort. When I introduced myself as a journalist who would be attending her ceremony, she poured out her story of shingles and unreceptive doctors and unrelenting pain.

"I've been suffering, suffering for five months now," she said, her voice wavering. "I can't get out of bed. My husband has to do everything for me." Her husband, frail and ancient, nodded soberly. "Tonight I'm going to ask the peyote to help me. I'm putting faith in the medicine to heal me."

I murmured sympathetically. I thought it unlikely that mescaline would do the trick.

Her wording captured the difference between Native Americans and a white girl like me. I thought of peyote as a chemical compound that alters consciousness; they thought of it as a divine gift throbbing with supernatural power. In my conversations with Mary Ann and others, I noticed that Navajos referred to peyote as a being, a personality. One called it the flesh of God, another called it a spirit. A third said it was holy medicine intended by the Creator only for native people.

Peyote, Andy Harvey explained, is "a sacrament."

With laugh lines around his eyes and mouth, Andy's face welcomed all strangers. His lustrous black hair took a decade off his forty-nine years. His stocky body suggested that he had not seen much physical labor since he began working for the economic development office of the Navajo Nation.

Peyote is like the Eucharist for a Catholic, he said. It *is* God, and it *opens the way* to God.

"We take it, we pray with it, and we ask the Creator to help us in the experiences that we would have as we ingest that peyote. So in a sense we ask the peyote to help us to come to realize certain things in our lives that we want to realize. And some things we can't understand, when we want answers to certain things. And in the case of being sick, sometimes we ask the peyote to help us cleanse the illnesses away and cleanse our mental being, our spiritual being, our psychological and emotional being. And we believe that's what peyote does. That's why we call it a sacrament."

"For a Christian, it might be like addressing Jesus," I offered, trying to translate into a metaphor I could understand. "It's like an intermediary between you and God."

"Right," he agreed. "And just like Jesus Christ was sent to save mankind, to provide us a means to salvation, this sacred herb has got the same philosophy. We believe that the Creator put it on this earth for us to utilize and provide salvation, and provide us a life that we could be satisfied with. And," he added quickly, "I've never taken it for the fun of it."

"The word *peyotl*," John Halpern explained to me a few minutes later, "means 'heart of God.' And these people believe it is a blessing of God to all native people, for their healing, for their strength."

Halpern and I were seated on a log a few hours before the ceremony, basking in the mountain sun. At thirty-seven, he was balding, enthusiastic, a study in kinetic energy, and very smart: Halpern was the associate director of substance abuse research at Harvard's McLean Hospital. Years earlier, he began studying why members of the Native American Church, who ingest peyote as religious ritual, enjoy vanishingly low rates of alcoholism compared with other Native Americans.[1] From there it was a short hop to fascination with all things Indian, including attending several peyote ceremonies. It was John who had wangled me an invitation to the ceremony, and for that I was grateful.

"They refer to peyote as medicine," Halpern continued, "but it would be with a capital *M* rather than a small *m*. We're talking about a *spiritual* Medicine."

Yes, but it's also a psychedelic drug, I pointed out.

He nodded. Peyote contains mescaline. It is a "phenethylamine hallucinogen" that has stimulant-like properties. Translation: it acts like speed and dramatically alters your state of consciousness, and if the dosage is high enough, you will experience visions like those described by William Blake and Saint Teresa of Ávila. That would be 350 to 400 milligrams for an adult, though how that translates into teaspoons from a coffee can, I could not guess.

Halpern explained that like more commonly known hallucinogens, such as psilocybin, LSD, and DMT, peyote targets the serotonin system in the brain, which is the system that regulates mood and emotion, among other things. Also like other psychedelics—and unlike, say, alcohol—peyote is believed to make people hyperperceptive, because it stimulates the brain cells and increases blood flow to the frontal lobes, or the "executive" part of the brain.

Then I voiced the question that haunts me: Does peyote swing open the door to a different layer of consciousness, where "God" can be known, or does it merely spark chemical reactions?

"There is this claim that the use of hallucinogens will increase magical thinking so you think there are more connections than reality would show," Halpern conceded. "And it's so nice to believe that science can explain everything. But do we really have to understand the machinations, the steps in which a person ends up communing with God? Is that going to serve a useful purpose to operationalize a human experience? To what end?"

I could almost hear his loyalties ripping down the middle as he gave his impassioned speech. John Halpern the Harvard doctor knew that the whole purpose of science is to "operationalize" human experience, to dissect it in order to understand it. And yet, he clung to the mystical. "We are spiritual beings," he said, "and to deny it does not make it go away."

Soon, I would see all these faces of peyote: the chemical and the god, the sacrament and the medicine—but a medicine with strings attached, one that required confession before it made you whole.

Confession and the Headless Man

"I need to confess to the fire," Mary Ann said, studying the flames as if to confirm they were ready to listen.

"One night, about twenty years ago, I was driving through Utah with my kids," she began. "It was the middle of the night, and we were coming over a hill. Suddenly, there was a man waving at me to stop, but we were going too fast. And then I saw a body lying in the road—he was already dead, but I ran over him, I couldn't stop. I think I ran over his head."

Startled and relieved that the endless drumming had ceased, I glanced around the circle to see if anyone had registered what may have been a confession to vehicular homicide. They merely smiled benignly, heads swaying happily like bobble-head dolls, and stared into the fire.

"I know I didn't kill him, but he kept haunting me," Mary Ann continued, a bit frantic now. "Once, a few years later, I was in my car, waiting outside a [peyote] ceremony. And a headless man jumped up outside my car window. I screamed. I was so scared. Sometimes I still see him."

She began to cry. "I need him to forgive me. I know I'm in pain because he hasn't forgiven me. I need peyote to heal me."

Mary Ann fell silent, and Fred, the "roadman" orchestrating the ceremony, began to chant prayers in Diné. The peyote man grabbed his tin full of paste and scooped a spoonful into Mary Ann's mouth, as a mother bird feeds worms to her chick . . . one scoop, two, three, four, five scoops, then a peyote button, washing it down with several gulps of peyote tea.

Mary Ann leaned back, sated. The drummers resumed their chanting. Two hours passed.

Two-thirty a.m.: "I had a vision," Mary Ann stated, "a vision of a bald eagle."

The others nodded. The peyote men fetched their medicine, and the group consumed the mescaline again, all except me. Then I noticed Mika, John Halpern's Japanese wife. She bowed forward as in a formal tea ceremony and, her face a mask of stoicism, rid herself of the sacred herb. No more peyote for Mika. Two more hours of drumming, and then, at four-thirty a.m., Mary Ann spoke again.

"The shingles are gone!" she cried triumphantly. "The peyote has healed me!"

Right, I thought, *let's revisit this issue in forty-eight hours.*

The next three and a half hours crawled by in comically slow motion. The ache in my legs escalated minute by minute, although this did inspire prayer: *Please, God, let this end.* I found myself gazing at the smoke hole at the top of the tipi, willing with all my might for the inky black sky to turn blue. But I was also mesmerized by the prayers and chants, and for the briefest of moments I did not want the ceremony to end. At eight a.m., eleven hours after we had crawled into the tipi, the roadman and his wife passed around water, corn, meat, and fruit as a ritual closing. The ceremony ended with a final blessing for Mary Ann: six spoonfuls of peyote paste, a button, and tea. *There goes that interview,* I thought as I watched her final peyote hurrah.

After the ceremony, a gauzy-eyed Mary Ann told me she was "too peyoted up" to talk with me, as did most others I tried to interview. After a few hours of waiting around, I reluctantly started off for Albuquerque without hearing Mary Ann's story.

Wending through the mountains of Navajo country, I reflected that the ceremony I had just witnessed was more parts Catholic Church than Berkeley party. The tipi fire, and the church candles. The sacrament as peyote, and as a wafer. Confession to the fire and the people, to Christ and the priest. The Diné chants, and the Gregorian. The offering of food, and the collection plate. And after the service, in each case, the breaking of bread.

Still, peyote's chemical makeup made me question how mystical the experiences can be. Knowing God through psychedelics. Seeing God through hallucinations. Hearing God through the serotonin system. Is it real? For Navajos, yes. Is the Eucharist the body of Christ? For Catholic believers, yes. Do I communicate with God through prayer? Yes, I believe I do, and when I am praying deeply, my brain waves no doubt slow, or my serotonin levels probably rise—and in that altered state of consciousness, I find God. Who am I to say that peyote does not unlock the door to the transcendent?

Two months after the peyote ceremony in Lukachukai, I called Mary Ann from my office in Washington. I was curious to know if her shingles had really disappeared, as she joyously proclaimed in the middle of the ceremony, or if the "healing" wore off with the peyote.

"Oh! That night, the pain stopped and never came back," Mary Ann sang happily.

"You're kidding. What happened?"

"While I was fixing the Medicine [the peyote] before the ceremony," she said, "I talked to the Medicine. And I said, 'I want You to help me. You know all the suffering I'm going through, and I want to get well with You. I want to put my faith in You, and trust You. Help me. Doctor me. Fix me up.' That's what I was saying to the Medicine while I was fixing it. So it did. It *sure* did."

"Remember how you talked about the man you hit," I said, recalling that Mary Ann had been raised Catholic. "Was that a confession?"

"Yup. You can't hide anything from peyote. You have to tell it, this is what the sickness is all about. This is where the sickness comes from. So the main thing about the shingles was the fear. The person that I ran over. It wasn't my fault. Someone else ran over this person. I didn't see it and I came up the hill and I couldn't put on the brakes so fast or stop. I got so scared, and I never went back."

"So the peyote cleansed you when you spoke to it?"

"Uh-huh. So when I prayed with the spiritual food, the person [who had been run over] came in front of me, sat in front of me. And I told him, 'I don't know who you are. I know you have a mom and dad. And I know you have grandparents. And I know I ran over you. But it wasn't my fault. You have to forgive me, it wasn't my fault. And because of that, I'm sick. I want to get well. I want you to forgive me.' So the spirit just left me. He came in front of me, and he left me. And that's when my pain stopped. He forgave me."

She giggled girlishly. "I can't believe I'm alive again! I'm so happy. The pain never came back at all. It was a miracle."

We hung up then, and I considered the possible explanations. The first, preferred by hard-core materialists, is that peyote has antiviral properties that can cure the shingles. Many doctors and psychologists would no doubt opt for the second choice: that Mary Ann's pain was in part psychosomatic, and once the fear and stress had dissipated, so did the pain. But spiritual believers hold firmly to the mystical explanation: Mary Ann had enjoyed a genuine spiritual healing, empowered by confession and forgiveness, by the supernatural character of peyote, and by prayer. Of course, she could simply be lying, but that seemed the least plausible of all the explanations.

Peyote has been intertwined with the divine for centuries among native people in North and South America. But only in the past two decades have scientists possessed the technology to peer into the brain and witness the effect of the sacred herb. Now, in the twenty-first century, drugs allow neurologists to watch the mystical experience unfold *in real time*. They can view the hand of God, or God's chemical proxy, as it courses through the brain, stimulating some parts, damping down others, to transport the subject to heaven or hell. It is like having God on tap.

For nearly four decades, this synthetic God lay just beyond their grasp. The war on drugs outlawed both recreational use and scientific research on the effect of psychedelics in the 1970s. It would take thirty-five years for the government to relax its grip on this scientific research. But happily, in 2006, more than 2,000 miles away from that tipi in Lukachukai, a neuroscientist at Johns Hopkins University had just published a study on the chemistry of mysticism.

God's Chemical of Choice

Roland Griffiths led me through the institutional halls of Johns Hopkins University Medical Center, turned the knob on a nondescript door, and ushered me into his mushroom mecca. The plush carpet sank beneath my feet, the pastels in the room soothed my eyes. I spotted a statue of the Buddha, a Shinto shrine, a crucifix, a painting of a vast landscape, something for every religious sensibility or none at all. Here, thirty-six people—one at a time over the course of many weeks—had rested for six to eight hours under the influence of thirty milligrams of psilocybin. They had stretched out with eye-shades and headphones attached to a stereo, cutting off the normal sensory information so that the psilocybin could hold court in their minds.

Griffiths motioned me toward a deep chair. He moved toward a white sofa, where he folded his thin legs into the lotus position. He was the chief investigator in a Johns Hopkins study on psilocybin and spiritual experience.[2] He was older than I anticipated, entirely gray-haired and drawn. A self-confessed "gym rat," he was strikingly slim in the style of people too busy to eat.

This breakthrough project, Griffiths confessed, sprang from a metaphysical question that dogged him for years.

"I started meditating ten years ago, and it opened up a window to the spiritual world," he began tentatively, gauging my reaction.

I nodded.

"It was astonishing. I felt as though I had connected to another reality. Somehow, we are pre-wired to have that experience—penetrate that reality. But why? Evolutionary biology? Or is there some sort of mysterious pointer? If we're wired that way, I want to know *who did the wiring*!"

I gazed at him, reflecting that most of the scientists I met who researched spirituality had experienced something mystical themselves. Personal, urgent curiosity—*What happened to me?*—drove them to the kind of alternative research that was at best a detour from the mainstream and at worst a risk to their careers. I sympathized completely. That curiosity drove me as well.

Griffiths continued. "And I thought, here I am a full professor at

Johns Hopkins, flying around the world presenting papers on drugs, when there are huge, much more compelling questions to explore. What is the nature of consciousness? And finally, we got some modest funding to study spirituality and what happens to the brain in an altered state of consciousness. We're just beginning, but it's been the most exciting research of my career."

With that buildup, you might think Roland Griffiths had extracted the God serum, a distinctive juice that ignites the brain into orgasmic mystical experience. In fact, his findings were somewhat more modest. This is the headline: Psychedelics can spark mystical experience. More precisely, twenty-two of the thirty-six volunteers (more than 60 percent) reported full-blown mystical experiences. They described feelings, visions, and insights that appeared to be similar if not identical to those experienced by mystics down through the centuries.[3] And, like other "natural" mystics I had encountered, Griffiths's subjects saw their worldview, their relationships, and their priorities rearranged by the experience, which they considered as meaningful as, say, giving birth to their first child.

Hardly the discovery of the double helix, surely, but allow me to inject a little perspective here. Science usually measures its progress in millimeters, not miles. In the world of neuroscience, which has endured a deep freeze on studying psychedelics, the fact that the government gave Griffiths permission to administer psychedelic drugs to volunteers *at all* was a watershed. The camel's nose was in the tent, and soon more researchers would follow with similar studies.

As neuroscientist Solomon Snyder, the lean, stately chairman of the neuroscience department at Johns Hopkins University, put it: "It's not what the dog says that's important. It's the fact that a dog can talk at all."

But for me, what the dog said deserves attention in itself, because the dog is talking about *serotonin*. Remember, we are looking for a "God chemical" that opens a person's mind to another dimension of reality. And psilocybin—the psychedelic given to Griffiths's volunteers and to my happy compatriots in the peyote ceremony—affects the serotonin system.

I asked Snyder whether Griffiths had caught the scent of a "God chemical." He considered the question, then responded carefully.

"Seeking the locus of religion in the brain," he said, "is by no means fanciful."

Snyder clearly does not traffic in hyperbole, so let me elaborate on his response a bit: What he is saying is that neuroscientists may be able to find the divine spark in our heads. Is that big enough?

Snyder himself became intrigued by the serotonin system when he was a medical intern in San Francisco in 1963. It all began when a friend lost her job at the medical center at the University of California in San Francisco. As a parting gesture, she filched a bottle of LSD from the laboratory refrigerator. For the next few weeks, Snyder and his friends gathered together every Sunday to do "research."

Those Sunday trips led this young agnostic Jew to a religious insight. Thanks to a glass of LSD solution, he could see through the eyes of mystics.

"I had this sense of being at one with the universe, which involves a loss of ego boundaries, such that you can't tell the difference between yourself and the rest of the universe," he said, looking slightly embarrassed. "Now, that doesn't make any sense. But let me tell you, that's what it feels like. And there's a vast literature, where everyone who has a mystical experience—whether Christian mystics, Zen mystics, or whatever—they describe almost exactly the same thing."

Snyder told me that scientists have long suspected that spirituality—or at least the farther edges of it, such as mysticism, hearing God, feeling the presence of the "Other"—might have something to do with neurotransmitters in the brain. One candidate is dopamine, the feel-good chemical involved in Ecstasy and "runner's high," although dopamine has received less attention in this context. Most scientists have zeroed in on serotonin and the serotonin system as the main triggers of transcendent experience.

Millions of Americans have an intimate relationship with serotonin: Prozac and other antidepressants are believed to smooth out moods by increasing the activity of serotonin in the brain. It turns out that LSD, psilocybin, peyote, and several other psychedelic drugs look very much to the brain like serotonin. When psilocybin enters the brain, it glances around for a place to camp out, and heads toward some very specific docking stations called serotonin 5-HT2A receptors. This is a close cousin to the serotonin receptor that Swedish doctors identified when they were looking for a "God gene," or a genetic predisposition to spirituality.[4]

This fake serotonin is an unruly thing. Imagine a bunch of six-year-olds being let out of school, their mothers waiting in the playground. They burst through the door and make a beeline to their mothers, who give them a hug and take them home. One day, a bunch of imposter six-year-olds, virtually identical to the real children, run to the unwitting mothers, who then hug them and take them home. Those imposters are psilocybin, masquerading as serotonin. The psilocybin excites the receptors and confuses them—or, in our analogy, the imposter children wreak havoc once they get home, upending furniture, knocking over lamps, smearing their chocolate-covered fingers on the walls. In the brain, the psychedelic drug creates visual and auditory perceptual changes, a sense of boundlessness, and the loss of time and space.

Having God on Tap

It is one thing to pinpoint the chemical or chemical system that reveals "God." Now some researchers, like Franz Vollenweider, are exploring *how* a chemical acts on the brain to foment spiritual experience. Vollenweider is a neuroscientist at the University of Zurich who has made his life's work understanding the chemistry of emotion and, more recently, spiritual experience. Happily for the science of spirituality, he lives in Switzerland, which has long permitted research using psychedelic drugs. That means that Vollenweider has accomplished what American researchers could not: he has observed synthetically induced mystical experience in real time.

Spiritual experiences are slippery little fellows; they generally do not occur when someone is lying in a brain scanner, being checked for a brain tumor. The only reliable way to study the neurology of spiritual experience is to actually trigger a mystical experience—that is, administer a drug to a subject, slip him into a brain scanner, and then witness the mystical experience as it unfolds. In this way, Vollenweider can scrutinize the physiology of the "God experience" as easily as other scientists can monitor sleep cycles or an epileptic seizure.

Vollenweider's research has identified serotonin, and a particular serotonin receptor, as keys to mystical experience. When the psychedelic drug psilocybin enters the brain, he told me, it makes a beeline for the serotonin receptor 5-HT2A, the same one that Roland Griffiths had targeted.

"When we blocked that receptor, nobody has any [mystical] experience," Vollenweider said. "It's the starting point. It's the major docking station."

"So serotonin is the God chemical?" I asked.

"Yeah, one of them," he said, laughing. He noted that other neurotransmitters, such as dopamine and glutamate, contribute to mystical experience and visions. But the serotonin receptor is a little like a bouncer at a party: if the psychedelic drug cannot get past that serotonin receptor, it cannot join the fun.

Once the drug passes the serotonin bouncer at the door, the party takes off, with brain chemicals interacting like dancers at a nightclub, bumping and grinding and creating a cascade of other reactions in the brain. Vollenweider has analyzed those chemical reactions and, in the process, he believes he may have located places described in the Bible and *Paradise Lost*: heaven and hell, and even the parts of the brain that spark the visions associated with those biblical "locations."

In trying to map this mystical topography, Vollenweider picked up the mantle of Adolf Dittrich, a famous German psychologist. In the 1970s, Dittrich tested hundreds of subjects and found that people tended toward one of three states when their consciousness had been altered by drugs, meditation, fasting, hypnosis, or other techniques. They experienced heaven ("oceanic boundlessness"), hell ("anxious ego dissolution"), or mystical visions ("visionary restructuralization"). Vollenweider's arrival at the Zurich Medical School in the 1990s marked a quantum leap in the research. He combined Dittrich's theory with brain-scanning technology.

"I thought, Okay, if these dimensions really exist, there must be a common correlate of these dimensions," he told me. "I began in 1994 and 1995, and I tried different inducers like amphetamines, ketamine, psilocybin, and MDMA." (MDMA is the drug known popularly as Ecstasy.) Vollenweider asked volunteers to take these drugs and then lie in a brain scanner for a PET scan.

"And we found that different networks *do* correlate with different kinds of experience," he said. In other words, one combination of brain activity induces hellish experiences, another confers bliss, and still another combination stirs up visions.

Heaven, Hell, and Chemicals in the Brain

Let's look a little closer at these states. Moving from the abstract to the empirical, consider the case of Michael Hughes. Mike had applied to participate in the Johns Hopkins study, but he did not qualify because he had (lots of) experience with psychedelics. Mike told me he had been profoundly changed by a spontaneous mystical experience as a teenager, and that had converted him into a perennial spiritual seeker.

Mike exuded that peculiar calm of so many others I have interviewed who have encountered the mystical. When Mike and I met in Baltimore on a sticky September day, he was sporting a faded blue polo shirt, blue jeans, and a big mop of brown hair with long, graying sideburns. Mike told me he had journeyed on a range of trips and would be my Virgil, guiding me through heaven, hell, and visions.

We'll begin with hell.

The savage nightmare stemmed from a tiny mistake. Mike was in his early thirties, staying with some friends in a cabin in West Virginia for the weekend. Mike measured out what he thought was a reasonable amount of dried, powdered mushrooms (psilocybin) for himself and his friends. He miscalculated.

"About twenty minutes into the experience, my friends and I could barely communicate with each other," he recalled. He looked stricken as he summoned the memory. "We were laughing, but the laughter was tinged with an 'Oh, shit, we've really done it this time' sense of foreboding."

Mike wandered off to another cabin on the property and crawled into bed. His body felt thick as an oak tree. He pulled the covers over his head, but could not bar the images from his imagination.

"It was chaotic and phantasmagoric," he recalled. "Suddenly, I felt as if something evil and malignant had entered my head—an entity of some sort. I had the sensation of tentacles moving around inside my skull. It was terrifying. I started praying the Lord's Prayer, over and over, like a verbal talisman. Though I'm not religious, I thought maybe just praying, saying anything over and over, would get the thing out of my head. I felt the tentacles probing and moving around, slithering. It was utterly horrific. I wondered if the thing, whatever it was, was going to suck out my mind."

Just as suddenly, the tentacles withdrew. "I was rolling around in the bed, sweating, confused, and panicky, but glad I was still sane."

If Franz Vollenweider had been sitting by Mike's side at that moment, he may have patted his hand and consoled him: *Now, now, you're just experiencing anxious ego dissolution, nothing to be afraid of.* According to Vollenweider, bad trips arise from an overactive thalamus, the little gate that filters sensory information. Vollenweider speculates that during trips gone awry, the thalamus lets in too much sensory information, too many lights, too many voices, too many visions. In the nightclub analogy, there is no crowd control; a bunch of bad characters gets into the room, and that leads to sensory overload, anxiety, disordered thinking—Mike's paranoia about the tentacles in his head threatening to suck out his mind. In fact, an overactive thalamus is associated with schizophrenia.

Most of Mike's experiences, however, took him to Vollenweider's heaven. One journey makes his voice quaver with awe to this day. He was twenty-two years old on that early summer night at St. Mary's College in southern Maryland. He and his friend, just finished with college exams, celebrated with some mushrooms. They wandered around the town for a few minutes, eventually finding the door to a historic Catholic church unlocked.

"I had never really been in a church by myself at night and was surprised to see that the candles were lit. It was almost as if I had wandered into this magical place," Mike recalled. "I sat down and I felt a really strong sense of sacredness. It felt like the accumulated energies and prayers of all the people who had been in this historic church for a hundred years had sort of congealed, that the atmosphere felt very thick with the presence of many, many people. I sat back and closed my eyes and I was overcome with this really profound sense of goodness and rightness and that everything that was, no matter how we felt—good or bad—was just as it was supposed to be," he said.

Hearing this, I nodded. I recalled Arjun Patel using the same words during his spontaneous mystical experience. *There wasn't any "me" any-*

more, he had told me. *There was just total seamlessness.* And did that feel good or bad? I had asked. *It just felt like . . . this is the way it's supposed to be.*

Mike continued, snapping me back to the present. "And it was really overwhelming. Trying to describe a transcendent experience is difficult. There aren't the proper words in our vocabulary to describe this immense sense of connection with *Something.* It felt like an intelligence to be sure, but it felt like a natural intelligence that imbues everything. It imbued the wood of the pews. It was emanating from the candle flames. It was the memories and the prayers of all the people that had ever been in this church. And it really felt like a microcosm of the universe itself. And at that point I really felt central and tied into all of existence."

Here is the "oceanic boundlessness" of Franz Vollenweider's heaven—the serenity, the unity with all things, past, present, and future. Here is the deep meditation of a Buddhist monk, the joyful vision of a Catholic mystic. In this state, you are flooded with happiness and peace. You feel no boundaries and are at one with the universe. When Vollenweider scanned people's brains, he found they experienced heaven when the chemicals stimulated the front of the brain. The frontal lobes keep you alert and processing information. At the same time, Vollenweider found that during good drug trips, the parietal lobes, which help you perceive personal boundaries (where your body ends and the rest of the world begins), were also stimulated.[5] Meanwhile, the amygdala—the part of the brain that processes emotions, anger, and fear—is dozing. The result is a mellow, blissful party.

As Mike narrated his story, I realized his "heaven" was woven together with visions, or, as Vollenweider would have it, "visionary restructuralizations."

"The candles were extra-radiant," Mike continued. "The colors were very rich. The shadows were very dense. These sorts of substances do bring out elements of our perception that we don't normally access.

Some people would say it's a hallucination, whereas I don't believe that. I tend to agree with Aldous Huxley that our perception is actually widened, and not just altered or twisted. It was also very tactile in the way my body felt opened up as if every pore was expanded, and it's hard to say whether the experience was more physical or mental or emotional, because it was so profoundly all of those at one time."[6]

If the Swiss neuroscientist had been watching Mike's brain in the scanner, he might explain that visions (or "visionary restructuralizations") refer to altered perception—hearing voices, seeing colors more brightly, even hallucinating.[7] Visions materialize, Vollenweider speculates, when the drug stimulates the striatum. That is a part of the brain that processes sights, sounds, tastes, touch; stimulating this area makes those senses richer, more acute, more colorful—but not so much that it terrifies. Ordinary things seem remarkable. *Oh, wow, look at that doorknob.*

Dave Nichols, a pharmacologist at Purdue University, speculated that Mike's brain was experiencing a torrent of events as he gazed at the candles. The part of the brain that detects colors was stimulated, and the frontal cortex, which makes sense of the world, was in overdrive, overprocessing the colors and making them appear richer. At the same time, Nichols believed, a part of the brain called the locus coeruleus was firing out signals in bursts. This part of the brain is nicknamed the "novelty detector"—because it tells you, "Oh! Someone just walked into the room! Pay attention."

"Say someone takes psilocybin and they look at a flower, and they say, 'Oh my God, this is really beautiful,'" Nichols explained. On the one hand, the signals in your brain that say "red rose" are now being processed in overdrive, so the red looks much brighter and more saturated. At the same time, he said, the novelty detector is going overboard.

"The normal connotation placed on the flower is, 'Well, that's a flower, and I've seen a million of them in my life and it's just a flower.'

But now you've shut off some of the processing that would come through and tell you that. And you're also having this novelty detector saying, 'Wow, this is a really interesting flower. Look at all those veins and petals.' "

The Brain as Spiritual Filter

As I worked through the chemistry of psychedelic experiences, one question nagged at me: Is a chemically induced experience a *real* spiritual experience?

Honestly, I haven't the foggiest idea whether drugs trigger a genuine encounter with "God" or another reality. Nor does anyone else, incidentally, since that would require knowing whether or not another reality exists. It is possible that drugs are the bullet train to God. It is equally possible that this sort of spirituality is nothing but a figment of a chemically addled brain, a synthetic light show, a cheap imitation of the deep-seated spirituality gleaned from prayer, or meditation, or life events that lead to epiphanies.

Arjun Patel takes the latter view. Arjun's appetite for the supernatural was whetted when he enjoyed a spontaneous mystical experience while meditating in his college dorm room. When he could not generate a repeat performance through meditation, he dabbled with mushrooms.

"Maybe if I hadn't had a spontaneous mystical experience, I would have considered some of my psilocybin experiences to be mystical," he told me. "But they're really very different. With psilocybin you know you've taken something. There's a physical, somatic component to it. I remember the first couple of times I tried it, I thought, 'Wow, I've just eaten a poisonous mushroom and this is the side effect.' "

And yet, others who have experienced mysticism through chemistry swear it is as genuine as the natural kind. Mike Hughes insists that his mushroom trip in the Catholic church was just as mystical as the spontaneous experience that launched him on his spiritual journey several years earlier.

"The two experiences were of equal profundity," Mike told me. "That's another thing I've thought about these years when listening to the debate: 'Oh, a chemical cannot make you have an experience of God.' I don't believe that. And ultimately, I don't really care if it is my brain chemistry doing this. They were equally profound. They both changed me dramatically. They're just two different roads to the same place."

I secretly hope Arjun Patel is correct. It peeves me to think that some people can ignite their faith life with a pill, probably because I lack the courage to do so. It irks me to contemplate all those hours of study and prayer, down on my knees straining to connect with God, when I could have swallowed a mushroom. But these irritations pale next to other puzzles: Does a chemical take one to a spiritual realm or to a subconscious one? Does it transport one to the light outside Plato's cave or into the folds of one's brain? And is "God" the creator of all life, or the creation of a chemical reaction?

Enter Aldous Huxley, who hatched what I consider the most helpful analogy for explaining spiritual perception. Best known for his novel *Brave New World*, Huxley was an early advocate of LSD—not for the thrills but for the insights it could bestow.

In *The Doors of Perception*, Huxley proposed that the brain is a "reducing valve."[8] He suggested that all around us is what he called "Mind at Large." This Mind, Huxley ventured, comprises everything—all of reality, all ideas, all images, seen and unseen, in the universe. The closest analogy to Mind at Large would be the Internet. Each of us has access to that information, but an ordinary fellow, or even an extraordinary Einstein, could not possibly process the flood of information.

Therefore the brain acts as a "reducing valve." It narrows the information to a "measly trickle," only that information necessary for survival. And so we ignore thoughts of the cosmos to focus on the lion crouching behind the bush. We turn from the stirrings of transcendence to the e-mail on the screen. We nudge aside insights about the universe in favor of dinner. Most of us live our lives on that level of reality, satisfied we are missing nothing.

What drugs do, Huxley suggested, is temporarily open the valve, loosening the filter so that extraordinary perceptions are admitted.

"As Mind at Large seeps past the no longer watertight valve, all kinds of biologically useless things start to happen," he wrote. "In some cases there may be extra-sensory perceptions. Other persons discover a world of visionary beauty. To others again is revealed the glory, the infinite value and meaningfulness of naked existence."[9]

And to others is revealed the face of God.

"It's not the drug that causes these experiences," explained Bill Richards, a psychologist and co-researcher in the recent Johns Hopkins study involving psilocybin. "You don't take psilocybin to get rid of your neuroses or to experience a transcendental experience. Rather, what psilocybin does is unlock a door. It gives access to many different states of consciousness, some of which are trivial, some of which are profound, some of which are crazy, and some of which are beautiful and creative."

Richards would guide me to another insight. It was a signpost that would point me back to Christian Science, the religion I had dismissed a decade earlier. Sometimes, a brush with "God" or another dimension of consciousness transforms a person physically. It is as if the mere possibility of a hidden reality realigns a person's body.

Researchers started to chase down this mysterious connection between psychedelic experience and healing a half-century ago. They envisioned a revolution in the treatment of autism, depression, terminal illness, and alcoholism. Many of the early studies were haphazard, bearing a sort of gee-whiz quality. They also yielded stunning anecdotal results. But before researchers could compile convincing evidence, the golden age of psychedelics came to an end.

Cancer in the Age of Psychedelics

By the 1970s, the U.S. government was fed up with the sixties drug culture and Timothy Leary's call to "turn on, tune in, drop out." Favoring a butcher's knife where a scalpel would have done nicely, it shut down the sale of all psychedelic drugs, not only for recreational purposes but for research as well. A rich area of psychopharmacology thus became off-limits, but not before a cohort of rock-star researchers at Harvard, Johns Hopkins, Chicago, and other research institutions conducted a raft of studies on the effect of psychedelics on a range of mental diseases and phobias.

In the mid-1960s, Eric Kast at Chicago Medical School stumbled on a surprising side effect of LSD—one that hinted at the symbiotic relationship among body, mind, and, dare I say, spirit. Kast found that after being treated with LSD, patients with end-stage cancer suffered less pain and "displayed a peculiar disregard for the gravity of their situations."[10] Study after study confirmed this. From 60 to 70 percent of the terminally ill patients enjoyed a dramatic turnaround in mood and outlook on life after tripping on LSD. Many saw their pain level drop precipitously.[11] The question was: Why? Did the chemicals reduce their pain directly—or indirectly, by reducing their stress and thus their pain? Or did these studies hint at another, spiritual, dimension?

I wanted to know the backstory to these psychedelic studies. I tracked down a few of the researchers, now approaching their seventies, who were too experienced, or too old, to worry about today's scientific reductionism. One of those researchers was Bill Richards, who helped conduct the Johns Hopkins study. In 2006, I met Richards at his home office, nestled in lush wooded land adjacent to a wilderness park in Baltimore. He sat in his cozy office on a brilliant July morning, a man with copper skin and a gray goatee and a half-smile that put me instantly at ease. A few minutes in his calm presence and I thought, I wouldn't mind taking hallucinogens from this man.

Why did psychedelics help the terminally ill? I asked.

The subjects traveled to "another level of consciousness. The deeper the mystical experience, the better the outcome," he said.

"There's a study I did with a hallucinogen called DPT," Richards later remembered. "We divided the cancer patients into two groups after the fact—those who had experienced mystical consciousness, and those who hadn't. And we compared those two groups, and those who did experience the mystical consciousness manifested more capacity for intimate contact, for example, as well as decreases in anxiety and depression."

"And pain, right?" I asked him.

"Reduced pain for some," he replied. "When we were working with LSD, we recorded decreased reports of pain. That was also true of the other substances. But part of it was, the *meaning* of pain changed. Before, pain was in the center of the field of consciousness. People would say, 'I'm suffering, I'm scared, I'm in pain.' Whereas after [the LSD] people would say, 'Oh! The pain is still there but it's off on the periphery of consciousness.' And at the center of consciousness would be relationships with important people."

It occurred to me there might be something more spiritual at work in these terminally ill patients than merely rearranging priorities and appreciating the precious time they had left. I voiced that suspicion to Stanislav Grof. Grof had headed psychedelic research at the Maryland Psychiatric Research Center, before the research was shut down. In his fifty years of studying psychedelics and "nonordinary states," he has sat in on more than 4,000 psychedelic sessions.

Grof said the mystical experiences were in a class by themselves, because they altered the patients' concept of reality, an effect that neither antibiotics nor Percocet has achieved. I thought back to Mary Ann and the peyote ceremony. Aided by mescaline that night in the tipi, the Navajo woman had traveled to what she believed was another spiritual dimension, and when she returned, the pain had evaporated. What Mary Ann and Grof's subjects shared was mystical experience.

"They lost fear of death," Grof continued. "It's also something we know happens to people who have near-death experiences. They've been in a car accident or have cardiac arrest, and they come back and they say they're not afraid of death anymore. And we found out it had tremendous impact on pain. It frequently helped with pain that was not responding to narcotics. Or the effect was beyond the pharmacological effect of the drug. The relief sometimes lasted several weeks."

In his book *The Ultimate Journey*,[12] Grof offers vignettes of cancer patients who had taken psychedelics for his research at the Maryland Psychiatric Research Center. Their mystical experiences could have been lifted from the pages of Raymond Moody's *Life After Life* or other books on near-death experiences: the visions included hell, judgment, and—always—light and redemption. The patients emerged from the trips convinced that life and love extended beyond the grave. Often, Grof reported, the pain levels dropped so dramatically that the bedridden patients were able to return to work for weeks or months.

In a groundbreaking study of patients with terminal cancer,[13] Grof, Richards, and other scientists reported the case of a woman whose breast cancer had metastasized to her spine. When the doctors first met "Mrs. G," she was paralyzed from the waist down, anxious, and depressed. After her first LSD session, she emerged determined to work with her physical therapist, and after a few months, she was able to get around with a walker. But a year later, she learned her cancer had spread throughout her body and that she would soon die. She fell into a depression and received another LSD treatment. During this session, "the patient had the experience of passing through a series of blue curtains or veils," the researchers reported. "On the other side, she felt as if she were a bird in the sky soaring through the air."[14]

Mrs. G's pain eased dramatically. She was able to walk down the aisle at her daughter's wedding without so much as a cane, and amazed the guests by dancing with her husband during the reception. Six months later, she was considering returning to work, and asked for another LSD session. This time, she enjoyed a full-blown out-of-body experience. The session began smoothly, but Mrs. G became frightened when she saw a huge wall of flames. After encouragement from the attending therapist, she was able to pass through the middle of the flames, and at this point experienced "positive ego transcendence."

"She felt that she had left her body, was in another world, and was in the presence of God, who seemed symbolized by a huge diamond-shaped iridescent Presence," Grof reported. "She did not see Him as a person, but knew He was there. The feeling was one of awe and reverence, and she was filled with a sense of peace and freedom. Because she was free from her body, she felt no pain at all."

Shortly thereafter, Mrs. G was discharged from the hospital "in good spirits." Later, when she was troubled with pain, she could push the pain out of her mind by reliving her out-of-body LSD experience. A few months later, the cancer finally claimed her life—or, perhaps I should say, claimed her body.

Because of her mystical experiences, she had lost her fear of death, which the doctors believed dramatically reduced her pain for nearly two years after a bedridden, paralyzed Mrs. G first contacted Stanislav Grof and Bill Richards.[15] In trying to explain the dramatic recoveries of Mrs. G. and more than forty other people in the study, the scientists could have copied a page from a book by Carlos Castaneda.

"Some of the patients who experienced the shattering phenomenon of death and rebirth followed by an experience of cosmic unity seemed to show a radical and lasting change in their fundamental concepts of man's relation to the universe," the scientists wrote. "Death, instead of being seen as the ultimate end of everything and a step into nothingness, appeared suddenly as a transition into a different type of existence."[16]

This research, happily, is getting a second life. Two years after I interviewed Roland Griffiths at Johns Hopkins, he wrote to tell me that he and his team of researchers were soliciting volunteers for a new study. They want to treat cancer patients with psilocybin in a "scientific study of self-exploration and personal meaning"—picking up where Grof and Richards left off. Richards himself is clinical director in the new research.

By this point in my research, these eerie descriptions no longer startled me, so often had I heard them from spiritual people like Sophy Burnham and Arjun Patel, from broken and restored people like Alicia, from cherished people like my mother. These descriptions stitched together disparate experiences like quilting thread. The quilt would expand to include people who lived with epilepsy, those who had a brush with death, and still others who meditated for hours on end. It seems to me that this makes the challenge for an atheist scientist that much greater: instead of reducing transcendent experiences to mere chemical firings, he must produce a plausible explanation for all sorts of experiences that have no connection to one another.

What Science Has Established

The cancer studies dogged me. I mused about them constantly. Perhaps I heard echoes from my Christian Science past, which argued that the very act of touching the divine through prayer has physical consequences. I heard, too, the refrain I had learned as a child in a Christian Science household: *Your thinking is your experience.* How you perceive the world affects how you experience it. And let's be clear: this is *not* a metaphor about turning lemons into lemonade. It is an ontological statement. Your thoughts and prayers affect your physical reality, I was taught, because the spiritual world shapes the human.

I had assumed that my friendship with Tylenol and the world of pharmacology had weaned me from such ideas, but here they were, trotting back and demanding attention. They were joined by Grof's terminally ill patients and my Navajo woman, who had, with a little help from their friends, replaced one vision of reality with another. For these people, the blotches on the paper had turned into a gaggle of flying geese, and the world would never look the same. They had accessed something—call it a *spiritual* state or call it an *altered* one—and returned transformed. Perhaps, I thought, a spiritual reality had broken into their physical lives.

Thanks to technology, neurologists can now watch the mechanics of the most profound moments of one's life, including mystical experience. So, what has science established so far? It has confirmed that brain activity correlates to one's (spiritual) experience. In all likelihood, when Saint Paul or Sophy Burnham enjoyed their spontaneous mystical visions, certain neurotransmitters were coursing through their brains, exciting this lobe and calming that one.

What this does not establish, in my opinion, is that mystical experience is nothing but brain chemistry. After all, if there were an "Other" who wanted to communicate with us, of course He or She or It would use the brain to do so, as opposed to, say, the left big toe. Of course God would use the chemistry in our brains to create visions.

God would also use something else: he would use electricity. For if there is a God who wired your brain, He is a master electrician. Your brain crackles with tiny electrical reactions between its different lobes, and some of those reactions spark a spiritual experience. Here we encounter one of the oldest of scientific puzzles in understanding mysticism: Is spiritual experience an electrical storm in the brain that afflicted great religious leaders and mystics down the centuries? Is it faulty wiring that leads to a sort of madness, or superior wiring that leads to spiritual insight?

Now that neuroscientists possess the technology to tackle the problem, they are looking for the answer in the "sacred disease."

CHAPTER 7

Searching for the God Spot

THE SUN WAS STILL HIGH in the northern Canadian sky when I arrived at Laurentian University early in the evening on July 8, 2006. I had traveled to the remote town of Sudbury, Ontario, to meet Dr. Michael Persinger, an American researcher who had gained some notoriety in neuroscience circles—and among journalists—for his experiments in spirituality. Several years earlier, he had produced the "God helmet," a reconstructed motorcycle helmet that was supposed to evoke mystical experiences in its wearer. According to Persinger, the helmet would use weak magnetic fields to stimulate parts of the brain—in particular, the temporal lobe. This, in theory, would evoke a "Sensed Presence," the feeling that a nonmaterial being was in the room. In other words, through the wonders of neuroscience, the helmet could summon counterfeit angels or demons on demand. I wanted to see if it would work its magic on me.

A bouncy brunette Ph.D. student named Linda St. Pierre greeted me at the research laboratory. She immediately sat me down and gave me a battery of tests to gauge my personality—and, in particular, to see whether I was prone to epilepsy, religiosity, or suggestibility. Questions like: Do you have a feeling that there is something more to life? Are you afraid of mice? Do people tell you that you blank out (a sign of epilepsy)? Do you believe in the second coming of Christ? And (my favorite): Have you been taken aboard a spaceship? As I was completing the tests, the man himself appeared.

Michael Persinger was ramrod slim, taut, with a puckish expression. He wore a dark blue three-piece suit, with a gold watch and chain tucked into his vest. I liked him immediately.

"I heard a rumor," I said, "that you wear a three-piece suit when you mow the lawn."

"True!" he admitted, seeming pleased that this eccentricity had made its way back to the States.

"Interesting. May I ask why?"

"For comfort. Three-piece suits are so versatile. I take off the jacket when it's hot, and put it on when I'm cold."

"How long have you been doing this?"

"Since I was in high school at least."

I had boned up on Persinger's theories about spiritual experience, and they boiled down to this: spiritual experience is a trick of the brain. It can be triggered by head injuries and brain dysfunctions such as epilepsy, by the earth's magnetic fields, and by machines like his "God helmet."

Persinger laid out his theory about how, precisely, the brain creates spiritual experience. It was like listening to Mr. Spock in a *Star Trek* episode—he peppered his theories with just enough acceptable science to make them plausible. The left hemisphere of the brain is associated with language, he explained, and thus the sense of "self." The right side is more involved with "affective emotional patterns," or feelings and sensations.

"When you stimulate the left hemisphere, you're aware of your 'self'—of you as an individual," Persinger explained. "So the question to ask is, What is the right-hemispheric equivalent of that left-hemispheric sense of self? And the answer is, the 'Sensed Presence': the feeling of another entity, of another sentient being that has emotional, meaningful, personally significant, and expansive temporal and spatial properties."

If you stimulate the right side of your brain in a certain way, he said, you sense someone nearby. In this way, Persinger claimed to create "the prototype of the God experience." Persinger claimed that fully 80 percent of the 2,000 or so subjects who have donned the God helmet report feeling a sensed presence, as well as dizziness, vibrations, spinning, and visions.

Even better, Persinger believed he had found the sweet spot for spiritual experience: the right temporal lobe of the brain. The temporal lobe, which runs along the side of your head near the ears, is involved with memory, emotions, and meaning, as well as hearing and language comprehension.

Why zero in on the right temporal lobe? I asked.

"Mystical experiences are in large part associated with temporal lobe function," he said. "The visual experiences, the hearing, knowing, vestibular [balance] effects, the smell."

Persinger explained that the temporal lobe (and, in his view, *not* the presence of God) explains why the mystics of old were said to smell fragrant flowers.

"In fact, their sweat emits it," he said. "Areas of the temporal lobe probably affect the metabolism in such a way that your sweat has a certain smell, and many of the classic mystics are often described as having a smell about them that is very fragrant, like roses, and all of this is tied to temporal lobe function."

I would learn much more about the temporal lobe in the coming months. For now, I was eager to test the God helmet.

At around six-thirty p.m., Michael Persinger led me into the "chamber." This was a small room with an overstuffed chair and ottoman, covered with what looked like an Indian-style saddle blanket. I settled into it and Linda St. Pierre began to affix eight electrodes to my head. Once the electrodes were properly connected, Persinger eased a motorcycle helmet with its own solenoids onto my head, and covered my eyes with goggles stuffed with napkins. It was stunningly low-budget. I half expected him to hand me a bong. When I was fully blinkered and feeling fantastically ridiculous, I heard him snapping pictures. Then, with a hermetic *schlupf*, the door sealed shut, and I was left, wired up and alone, in my soundproof room.

In order to record what happened while I was in the chamber, I had placed a tape recorder (with Persinger's permission, of course) in the control room where the neurologist and his assistant were manipulating the magnetic pulse passing over my scalp. The recorder picked up Persinger's comments to the assistant (which I could not hear in the soundproof chamber), as well as his comments to me when he hit a button and spoke through a speaker in the chamber. It also recorded my comments to him, since I was wired with a microphone that fed into the control room.

"Ms. Hagerty," I heard Dr. Persinger's soothing voice through the speaker on the wall. "Can you hear me? It is important that you relax. The changes will be extremely subtle, so you must let the experiences simply arise."

Then the experience began. I relaxed into the chair, and waited. Soon, I grew a little agitated. I could hear the helmet clicking as the magnetic patterns shifted, but I felt and saw nothing. Minute unfolded into uneventful minute, and I began to worry: What if nothing happens? What if I fall asleep in this smelly chair? I found myself straining to see patterns in my mind's eye, wondering when—or if—the Sensed Presence would show up.

About this time, in the next room, Dr. Persinger was also becoming agitated, and dictated his concerns into the tape recorder.

I should point out, he said, *that your EEG is quite fast right now. It is way above the level that is optimal. If your brain activity is too fast, the applied field will not drive the neurons to produce the first stages of the mystical experience—that is, the Sensed Presence. Obviously you can't hear me, since you're in a closed acoustic chamber, but if you could, I would be suggesting you relax much, much more.*

Unaware of Persinger's concerns, but keenly aware of my own, I suddenly recalled a scene from ten years earlier, a previous time when I had felt spiritual performance anxiety. I had been invited to the house of an acquaintance for a "Bible study." I arrived to find six others there surrounding Nancy, the guest of honor. Nancy, I was informed, was a modern-day Christian prophet: she was filled with the Holy Spirit, and she heard directly from God. She visited Washington from Texas every few weeks to hold a Bible study and tell other women what God was saying about their lives. I quickly noticed that most of Nancy's "prophecies" involved money—people becoming rich, owning large mansions, giving away millions of dollars for evangelism projects. Most of the Bible verses she quoted in her "study" promised abundance.

After about an hour of Nancy's preaching, she slapped her hands on her thighs and said, "Now let's hear from the Lord!" in the way someone might say, "Let's have some ice cream!" Everyone murmured assent, except for me, who felt a little bit queasy.

We huddled together and Nancy prayed for God to speak through each one of us. Then she announced, "Okay, I want each person to prophesy over another person. I will begin."

She began to prophesy about me. She "saw" me on a "jet plane with a man with a turban on—he owns the plane. Oh! He's very wealthy. And I see you getting off the plane in a very hot dusty country. Oh! And you are being greeted by hundreds of children, little brown children, little brown babies. You must be in Africa," she added. And that was it.

One by one, the others prophesied. My hands grew clammy; I surreptitiously glanced at the door, plotting my escape. Finally, everyone had spoken. It was my turn to inform Sheila about God's will for her life. Silence fell on the room as I struggled for something to say. The seconds stretched to the breaking point, the silence yawned wider and wider, and I saw . . . nothing. I didn't have a word for Sheila. Not one stupid word. I thought, *Oh, man, I'm drowning here.*

"I see water," I heard myself intone, in a flash of inspiration.

The others murmured, "Yes, Lord."

Warming to my tale, I said, "It is a pool, and you are standing on the high dive."

"Yes, Lord, thank you, Jesus," several exclaimed, and I heard someone softly speaking in tongues.

"You're afraid to dive, Sheila," I continued, "but the Lord does not want you to be afraid. The Lord says, 'I will catch you, my child, just trust me.' "

"Thank you, Lord!" This came from Sheila, in heartfelt gratitude for this, my first prophecy.

Six years later, I was at a reception at the Mayflower Hotel in Washington. A woman walked up to me.

"Barb?" she said, moving closer to examine my face. "Do you remember me? I'm Sheila. We met at Julia's house, when you prophesied over me."

"Of course, Sheila, how are you?" I said, scrambling to remember what, exactly, I had said.

"Great! Do you remember how you prophesied that the Lord wanted me to jump off the high dive?"

"Oh. Yeah."

"Well, the next day I quit my job!"

I wondered briefly if one could be sued for faking a prophecy. I was intensely relieved when she told me she had started her own animation and graphics company, which had prospered from the first day.

Back in Michael Persinger's chamber, I found myself once again expected to manufacture a spiritual experience. I willed myself to stop thinking and simply relax. As the minutes ticked by, a few images penetrated the darkness—a flash of a park near my house, the brief glimpse of a face. The most profound moment came as I tottered on the edge of sleep.

"I am utterly relaxed," I said in a thick, gravelly voice. "I feel as if I'm dissolving into the chair. It's not that I don't have boundaries, but the boundaries include the chair." Pause. "I'm communing with a chair." Pause. "Great."

For the remainder of the session, I felt blackness cresting over itself like roiling waves. I briefly saw a woman's face, but even in that suggestible moment, I knew these were creations of an imagination trying too hard, and certainly not the presence of a Sentient Being. Mercifully, the session ended a few moments later.

My answers on Persinger's follow-up questionnaire reflected my desultory performance. On only one question—"I felt relaxed"—did I give an enthusiastic response. I had also (erroneously, I came to believe) checked the item: "I felt I left my body," thinking about the union with the chair.

Dr. Persinger seized on that answer. I tried to explain that it was more relaxation, not an out-of-body experience, but he persisted.

"But you did feel something *like* this," he insisted, and in that instant I realized how he came by his remarkable results.

"Yes," I said, "but no Sensed Presence."

"Well, according to your EEGs, you were right on the verge of feeling a Sensed Presence," he assured me. The machine was showing very fast spiking behavior in my brain waves about three minutes before the session ended.

"If we had continued, it would have hit you"—he flicked his fingers in front of his face, as if releasing a ball of energy—"it would have hit you powerfully."

"Say it *had* hit me powerfully," I suggested. "Would that mean spiritual experience—God or Allah or whatever you want to call it—can be reduced to brain activity?"

"Well, certainly from the point of view of neuroscience, all experience is generated by brain function," Persinger said. "That means when you have an experience of a memory, that's a brain pattern being activated. When you have an experience of God, or Allah, or Buddha, or whatever the cosmic whole is that inspires you, that is brain activity. Does that mean that everything is programmed by brain structure or electromagnetic activity or chemistry? *Of course it is.*"

Simply because the brain is *firing* during these experiences does not mean the brain *causes* them, Persinger quickly added. But in the next breath he gave up the game, blithely stating what I suspect many scientists believe but do not say, at least not to religion reporters like me: Neuroscience will soon relegate "God" to the ash heap of history.

"If we look at the trend in the history of science, first we thought we were the center of the universe, and Copernicus modified that," he said. "Then Darwin removed the illusion that we were a special creation. Freud ripped apart the concept that we were logical animals and showed that that was only a veneer, that actually we were still a primitive animal.

"So the next big question is: What is the last illusion that we must overcome as a species?" he asked rhetorically. "That illusion is that 'God' is an absolute that exists independent of the human brain. That somehow we are in His or Her care. We have to realize that ultimately that may not be true, but what we may learn from it may take us much further than we ever imagined."

I left Persinger's laboratory near midnight, deflated. I had been secretly hoping for a spiritual experience, perhaps as confirmation that I had the right biological makeup, the right polymorphism in my DNA, to connect with a "Sentient Being." I guess I'm not nearly as spiritual as I want to be. Besides that, I was now highly dubious that Persinger's

God helmet could just tickle the temporal lobe and evoke a counterfeit God. Sure, his ideas were plucked from the work of pioneering neurologists like Wilder Penfield and cited by many neurologists today. But his research had drawn derisive criticism from other researchers.[1]

So imagine my bafflement six months later when I finally transcribed the tape that we had recorded in the control room. The recording revealed that Persinger's God helmet *had* in fact manipulated my temporal lobes and my experience. When Persinger began to stimulate my right temporal lobe, he predicted (into the tape recorder) that I would see slightly scary objects, such as faces, to my left. A couple of minutes later, my voice came through the speaker: *I see funny little goblins off to my left there, like we're in Sherwood Forest.*

Later Persinger predicted I would feel a sense of evil—once again, on my left, since he was stimulating the right side of my brain. About a minute later, my voice broke the silence: *I'm not sensing a presence per se, but there's kind of a roiling darkness, like a battle of darkness, like waves breaking over each other, but it's all dark, it's off to my left.*

Next Persinger narrated that he had changed the stimulation to a "burst firing pattern" to affect my amygdala, the part of the brain that triggers strong emotions, especially fear. As he was explaining this into the tape recorder, my voice interrupted: *I see little apparitions like little skulls.* A few seconds later: *I have these strange visual images, kind of over to my left. The face of a ghostly young woman, and then followed by a black goblin, with those eyes that you see in Japanese cartoon characters—spider-like eyes.*

Shortly after this, the battery died and the recording ended.

As I transcribed my words, I realized that in my skepticism, my memory had filtered out some of my responses to Persinger's electromagnetic waves. Manipulating my temporal lobes had evoked mental images—and, who knows, perhaps the memory of my prophecy for Sheila. I conceded that I could not easily dismiss Michael Persinger's work, nor his discomfiting assertion that God is all in one's head.

The Sacred Disease

In 400 B.C., Hippocrates wrote one of the first texts on epilepsy, "On the Sacred Disease." In his essay, the renowned physician of ancient Greece tackled the conventional wisdom of his day: namely, that epilepsy was a sacred disease bestowed by the gods. At the time, doctors believed it was a curse that threw its sufferer to the ground in paroxysmal fits; anyone who has seen a tonic-clonic (or grand mal) seizure knows it is a terrifying event to witness, and could look to a superstitious mind like demon possession. Alternatively, some ancient physicians believed epilepsy to be a divine blessing that conferred gifts of prophecy and spiritual visions. Not so Hippocrates. Hippocrates argued that epilepsy was brain dysfunction pure and simple, and that erratic brain activity, not the gods, sparked these fits.

Some 2,400 years later, the medical view of epilepsy has sided squarely with Hippocrates. It has vaulted from one extreme to the other—from superstition, where everything is attributed to divine whimsy, to reductionism, where everything is explained by brain chemicals and brain-wave activity. The vast majority of scientists now understand epilepsy to be a neurological disorder, a firestorm in the brain. Armed with evidence from brain scans and other technology, some modern neurologists are, like Michael Persinger, reframing religious history. The effect has been not to present the disease as sacred, but to present the sacred as a disease.

If scientific journals are to be believed, most of the world's religious doctrine sprang from dysfunctional minds. The list of religious leaders who, neurologists say, might have suffered from temporal lobe epilepsy is as long as it is impressive.[2] The most intriguing and controversial figure is Saul of Tarsus. Before he became Saint Paul, who authored much of Christian doctrine, he was a fiery rabbi who had set out on the road to Damascus intent on persecuting the early Christians. We pick up the story in Acts, chapter 9.

"And as he journeyed, he came near Damascus. And suddenly there shined round about him a light from heaven. And he fell to the earth, and heard a voice saying unto him, Saul, Saul, why persecutest thou me? And he said, Who art thou, Lord? And the Lord said, I am Jesus whom thou persecutest: It is hard for thee to kick against the pricks."[3]

Was this a conversation with the Son of God, or "visual and auditory hallucinations with photism and transient blindness," as the renowned psychiatrists Kenneth Dewhurst and A. W. Beard surmised?[4] Was it a divine appointment, or, as neurologist William G. Lennox stated, the emotional reaction to "the voice of conscience possibly complicated by a migraine-like syndrome"?[5]

Paul is not the only religious leader to have been retrofitted with epilepsy. Neurologists have speculated that Muhammad's visual and auditory "hallucinations" were complex partial seizures. The voice of the angel and the white light that inspired Joan of Arc, they say, did not come from heaven but from her daily ecstatic seizures. Joseph Smith's pillar of light and two angels—a vision that birthed the Church of Jesus Christ of Latter-day Saints (Mormonism)—may have boiled down to a complex partial seizure. Ditto for the mystics Teresa of Ávila and Thérèse of Lisieux, for Emanuel Swedenborg, who founded the New Jerusalem Church, and Ann Lee, who led the Shaker movement. Even Moses, who led the children of Israel through the wilderness and authored the Pentateuch, does not escape a troubling diagnosis. What, after all, are doctors to make of that encounter with the burning bush?

I called Orrin Devinsky, who directs the New York University epilepsy center, for his perspective. Part of the problem, he said, is that complex partial seizures, the most common form of adult epilepsy, were not medically understood until the mid-1800s. This particular parlor game involves people who lived scores or hundreds of years before that.

"So we're left with little shards, little fragments of descriptions,"

Devinsky said. "Moses at the burning bush. What was happening at the burning bush? He was looking at a bush that was burning but not consumed. Assuming a more rational scientific view, Moses was having a visual hallucination and then he heard 'God's' voice. As a physician, can I say that was consistent with a temporal lobe seizure? Yes. Can I say Moses had a temporal lobe seizure at that time? All I can say is it's a possibility. It also could have been a genuine religious experience. Or perhaps both."

For me, at least, a few things do not sit well with the saint-as-epileptic hypothesis. First, there is the imperious snobbery of scientific materialism, which implies that if a person's brain acts a little bit differently from the "average" brain—slow brain-wave activity or the occasional electrical surge—then anything that person believes is probably distorted, illegitimate, just a little bit crazy. This troubled psychologist William James as well. We would never think of dismissing van Gogh's art because in his madness he cut off his ear, James said, or throw Dostoyevsky's *The Brothers Karamazov* in the trash because its author suffered from epileptic seizures.

And yet, "medical materialism finishes up Saint Paul by calling his vision on the road to Damascus a discharging lesion of the occipital cortex, he being an epileptic," James observed in his famous Gifford Lectures in 1901. "It snuffs out Saint Teresa as an hysteric, Saint Francis of Assisi as an hereditary degenerate. George Fox's discontent with the shams of his age, and his pining for spiritual veracity, it treats as a symptom of a disordered colon. . . . And medical materialism then thinks that the spiritual authority of all such personages is successfully undermined."[6]

My second reservation about reducing spiritual experience to a seizure is this: epilepsy is a horribly debilitating disease, punctuating one's life with random fits of unconsciousness, short-term memory loss, and often eventual psychosis. It is hard to imagine epilepsy appearing frequently on the résumés of great religious leaders who traveled the Roman Empire with a new religion as did Paul, guided a nation in the wilderness for forty years as did Moses, or led troops into battle as did Muhammad.[7]

And yet, scientists have identified a connection between the temporal lobes and religiousness. Some say, only half joking, that the temporal lobe is a "God spot." While this mass of tissue is not the sole trafficker in spiritual experience—many other parts of the brain light up during transcendent moments—it appears to be a major player, necessary but not sufficient.

The God Spot

The search for the spiritual brain center began almost by accident, with the pioneering work of Wilder Penfield. In the 1940s and 1950s, the Canadian neurosurgeon began to excavate the brains of his patients before surgery to determine what parts could be removed (to stop the seizures) and what parts could be left in (so that he would not unnecessarily remove, say, the patient's ability to speak). Because the brain contains no pain receptors, Penfield could put his patients under local anesthesia and keep them conscious while he opened their skulls and poked around. He would stimulate various areas of the brain with an electrode and then ask the patient where he felt the prod in his body.[8] In this way, Penfield mapped the part of the brain that corresponded to the right thumb, or the lips, or the left big toe.

Penfield made a different sort of discovery when mucking around in the temporal lobes: a few of his patients reported out-of-body experiences, hearing voices, and seeing apparitions.[9] To be sure, these were nothing like the complex, hours-long visions of Teresa of Ávila or Joan of Arc. The sounds consisted of snatches of music or random words. The visuals included bizarre images or familiar scenes scooped up from their memory banks. Only two of Penfield's more than 1,100 patients reported anything like an out-of-body experience. Nor were any of their experiences awe-inspiring visits to paradise. One patient cried, "Oh God! I am leaving my body!"[10] And another patient said, "I have a queer sensation as if I am not here . . . as though I were half [gone] and half here."[11]

Still, Penfield's electrodes bore testimony to a suspicion that many doctors had harbored for decades—namely, that there is something special about the temporal lobes. *Could they be the seat of spirituality?*

Beginning in the 1960s, British psychiatrists Eliot Slater and A.W. Beard picked up the search. When they asked their psychiatric patients about their seizures, nearly four out of ten reported mystical delusions and hallucinations, including events like "seeing Christ come down from the sky," "seeing Heaven open," and hearing God speak.[12] Anecdotal evidence accumulated about the "sacred disease." For example, Dr. Beard and another colleague, Kenneth Dewhurst, theorized that people with temporal lobe epilepsy might be prone to have a religious conversion in the hours or days after a seizure. They described the cases of six people suffering from temporal lobe epilepsy. One woman heard "a church bell ring in her right ear; and the voice said, 'Thy Father hath made thee whole, Go in Peace!'" Another patient had a "day-time visual hallucination in which he saw angels playing with harps."[13] Something about their seizures triggered spiritual experiences in their brains.[14]

Although the spirituality-as-epilepsy theory remains controversial, with time, many neurologists settled into the idea that increased activity in the temporal lobes is central to spiritual experience.[15] If the evidence is thin, researchers speculate, that is because the clinical neurologists, the ones in the white coats who actually treat patients, generally do not ask about spirituality. *So, how is the Dilantin working for you? Okay. And your most recent seizure? Tuesday? And when was the last time you saw Christ descend from heaven?* For their part, patients hesitate to mention these types of visions; most already feel sufficiently self-conscious without adding the heavenly spheres into the mix.[16] Yet when I asked patients about their spiritual auras, the stories surged out of them with an urgency and a clear-eyed certainty that left me stunned and impressed.

Jordan Sinclair's Scar

There is no doubt in Jordan Sinclair's mind that he got religion from his temporal lobe.

"Of course I did," he told me. "I've seen the brain scans."

I met Jordan Sinclair (not his real name) through the Epilepsy Foundation, which let me post a note on its website. I wrote that I wanted to interview patients who had experienced ecstatic spiritual seizures or dramatic religious conversions as a result of their epilepsy. I received dozens of responses, but Jordan's was the most remarkable.

Jordan was raised in a Conservative Jewish home. His parents were both Holocaust survivors from Eastern Europe. Still, he never embraced religion.

"I always thought organized religion was a con game," he told me the first time we talked, in November 2006. "It always got me mad that on High Holy Days you had to buy tickets to pray."

Jordan constructed a life as a highly successful comedy writer in Los Angeles. In 2000, when he was forty-three, his doctor found and removed a benign tumor from his left temporal lobe. The surgery was a snap. But later Jordan began experiencing strange things. Sometimes when he was shaving, he said, his hands seemed unfamiliar, as if someone else's fingers were wielding the razor. One day while he was shopping at the nearby mall, the stores suddenly appeared unreal.

"The storefronts all looked like façades, like sets on a movie location. The lighting was wrong. The people—they looked real, but they kind of looked a little bit animated, a little bit of a Pixar quality to them. And I was so afraid for a moment there—it just looked like at any moment someone was going to come out and yell, 'Cut, we got it!' and tear away the entire mall, and there were going to be a million people working there, you know, cleaning up, and all the people were extras. That's how fake it seemed."

What Jordan did not know was that *jamais vu*—in which the familiar seems brand-new: the opposite of *déjà vu*—commonly plagues people with temporal lobe epilepsy. Shortly after the shopping-mall episode, Jordan visited his neurologist. His MRI films revealed that, as a result of his surgery, his left temporal lobe had retreated from his skull. It was a different shape and size, and new scarring was causing electrical storms in his head. Jordan had developed temporal lobe epilepsy. But the organic change in his brain was nothing compared with the one in his heart.

Jordan could pinpoint the moment of his spiritual epiphany. He and his eleven-year-old daughter were reading about world religions. They worked through the chapters on Judaism, Christianity, Islam, and Hinduism without event.

"When we got to the chapter on Buddhism, as soon as I started to read it, the light went on in my head, and it was almost as if I had found the definition that fit the things I was feeling and thinking."

He laughed, a little embarrassed. "I probably know less about Buddhism than you do. But it seems to me, Buddhism is the truth. I feel it. I mean, right now I'm looking at my fireplace mantel, where I have . . ." He inhaled deeply. "Let's see—one, two, three, four, five, six, seven, eight, nine, ten, *eleven* statues of Buddha looking at me. I'm wearing a *mala* bracelet, which is like a rosary. I've been wearing one every day—don't know what it means, don't know its significance, but I feel that if I'm not wearing it, I forget . . ."

His voice trailed off.

"I do know I've never believed more in spirituality, never believed more that it has answers to the most basic questions."

Neurologists might classify Jordan as having Geschwind syndrome, in which his small seizures rewrote his mental script, his values, and his thoughts, not only during the seizures, but all the moments in between. I told Jordan that scientists would probably relegate his spiritual conversion—the top-to-bottom overhaul in his values, his new Buddhist view that compassion trumps success—to a brain disorder.

"I know," he sighed. "It's kind of sad in a way to think that your religious beliefs, or your feelings of spirituality, or your peace and love for your fellow man, is a mistake that comes from an electrical impulse that's gone awry. But I'll tell you what the real bottom line is for me: *I don't care where it comes from.* I'm much happier feeling the way that I do now about spirituality and religion and tolerance and compassion. I'm a more decent human being every day because of it."

The Cosmic Electrician

Jordan Sinclair's story piqued my curiosity about how, precisely, a brain generates otherworldly beliefs. Or, for those who believe in God, I will put it this way: Is there a cosmic electrician who wires people's temporal lobes to let Him communicate with them—and if so, what does the wiring look like? Now some neurologists are exploring the same questions—zeroing in on epilepsy, not to *dismiss* spiritual experience as brain dysfunction, but to *understand* it. On the theory that the extreme can elucidate the commonplace, they are trying to decipher the mechanics of how a spiritual experience unfolds in the brain.

As Jeffrey Saver, a neurologist at the University of California, Los Angeles, put it, "These patients give us clues as to what parts of the human brain are involved when *all of us* have a numinous experience."

Studies suggest that when pollsters interview Americans in depth, some 60 percent report that they have had an experience of the presence of God or a "patterning" of events in their life that persuades them that they are part of a cosmic design.[17]

"We know that this is a universal response across cultures," Saver continued."And patients with these disorders give us an insight into the biologic hardwiring that underpins this universal human experience."

"So are you saying that if there were a God who wanted to communicate with us," I asked,"He or She would use the temporal lobes?"

"There have been two general hypotheses about how God might communicate with us," Saver explained, smoothly adopting my God terminology. "One is that there's some specific sensory organ that exists only to have contact with the divine. And so far we have no evidence that that sense organ exists. The other view is that we encounter the divine through our usual sensory faculties—through taste, smell, vision, our usual senses—but that when we encounter the divine, that contact is touched by, *stamped by*, some additional marker indicating that it is of special meaning, of special ultimacy. And the part of the human brain that stamps events as having these special qualities is the temporal limbic system. So that would be the way that a divine presence would manifest itself—if there were such a contact," he added hastily.

Let's look a bit closer at this part of the brain, and the "temporal limbic system." From the side, the brain looks like a boxing glove, and the temporal lobes are where the thumbs would be, covering the ears. The temporal lobes are the gateway for, among other things, ideas and emotions, particularly about yourself. Their brief includes hearing (remember, the lobes are close to the ears), language comprehension, conceptualizing, memory, and attaching meaning and significance to events. Within the temporal lobes is the limbic system. Think of it as a loop that performs a certain function, like the cooling system under the hood of a car. There are more than half a dozen regions involved in the limbic loop, but for our purposes, we will focus on two.

The hippocampus is the memory center, essential for the formation and storage of long-term memories. Next to the hippocampus is the amygdala, which serves as the fight-or-flight messenger: stimulate that area and you're likely to scamper away in terror or snarl in aggression. Working together, the hippocampus and the amygdala stamp people, places, and things with meaning. A woman who was bitten by a dog when she was a girl will shrink from my yellow Lab because her memory of dogs has been stamped with a particular emotional meaning (fear). But for someone like me, who grew up surrounded by dogs, my yellow Lab evokes only love and warm memories. Same dog, same parts of the brain, different meaning.

Now let's look at what happens when a person suffers a seizure. At the start of the seizure, the cells in the brain begin to move in rhythm with one another, creating a powerful synchrony. The symptoms the person exhibits depend on the area of the brain in which this rhythmical spiking occurs. If the seizure zeroes in on the motor area of the brain, you might see the right thumb twitch or the left arm jerk around. If the electrical storm occurs in the temporal lobe—which is likely, since the temporal lobe is the most electrically unstable part of the brain—then the person could experience an emotionally vivid aura—that is, the beginning of a seizure.

Because of the spiking, all these normal emotions have an exclamation point after them. They're scrawled in huge red capital letters. People may hear music or words, presumably from their memory bank—and instead of being just snatches of remembered sound, they are music from the heavenly spheres or messages from God. They may see flashes of light and interpret it as an angel visiting from heaven. They may feel a sense of unreality and disconnection from the world and believe that they are experiencing an alternate reality. They may sense the presence of another being and believe that presence is God, or the devil. They can feel terror or (in very rare cases) ecstasy. And these sights, sounds, and feelings have cosmic weight, because the storm in their brain is telling them so.

If this electrical storm rolls through often enough, it physically rewires the brain. And in this may lurk a clue as to why many people—Christian Pentecostals, Orthodox Jews, Sufi mystics—seem to look at the world through the prism of faith. It may be that their brains are tailor-made for God.

Tailor-made for God

On January 22, 2007, I drove into the sprawling complex of Henry Ford Hospital in Detroit. The day was bitterly cold, gray, with snow flurries threatening a storm that would soon drop a foot of snow on the city. Yet this was nothing compared with the neurological storms that I was about to witness. Dr. Gregory Barkley, the director of the epilepsy clinic, had graciously set up appointments for me to interview patients whose epileptic seizures had transcendent elements.

One of them was Mary. In the photo I took of her, Mary wears a royal-blue turtleneck and black pants. Her brown hair is streaked with gray and worn in a pageboy style reminiscent of the 1970s. Her head is cocked to the right, her narrow shoulders droop, and her lips press together in a sad, straight line.

The picture deceives. When Mary talks, she springs to life, a jack-in-the-box whose dramatic gestures sweep you into her sphere and shower you with a joy that belies her chronic disease. Mostly you feel her obsessive love for God. Very few sentences pass her lips without mention of the Blessed Sacrament or Jesus, the lives of ancient saints or present-day nuns.

Raised Catholic, Mary suffered her first tonic-clonic seizure when she was fifteen years old. With aunts who were nuns and a cousin who became a Jesuit priest, Mary was rich soil for a bumper crop of Catholic faith. I saw something urgent about Mary's spirituality: it saturated her life, it dominated her vision, it drove her relentlessly to share the Gospel. As she described her life, it seemed that every moment contained dual meaning—in this world and in the spiritual realm.

Nothing was coincidence. She recalled one afternoon when her seizure lasted only about thirty seconds. "And then I came to, and I said, 'Well thank you, Lord, I didn't even bite the sides of my mouth, Jesus!' And then a week later, I read that a man who had written a book on prayer that I had read, Father Ed Farrell, had died in his afternoon nap. I thought that was so neat, that maybe that was part of the repository of grace in the world. I believe in the mystical body of Christ, so maybe that [brief seizure] was required of me for his happy passage."

Or consider the day in 1985 when she was diagnosed with cancer.

"You want to know something neat?" Mary asked, leaning forward. "They operated on me on August 15, which is the Feast of the Assumption of the Blessed Virgin Mary! And I didn't pick that date. They did. And the first time I was seen by the surgeon afterwards was the first Saturday in May, and the month of May was dedicated to the Blessed Mother. And then my last appointment with the surgeon was on the feast of the Epiphany."

In recent years, researchers have searched for a reason why temporal lobe epilepsy would render a person hyperreligious. Early on, emotional reasons headed the list: the desire for solace from a loving God, for example, or a way to cope with and find meaning in a life punctuated with abrupt fits. But now neurologists are finding physical clues as well. Recently, for example, some researchers found that in people with temporal lobe epilepsy the hippocampus (the memory region) is smaller than normal.[18] Other studies showed that the amygdala (the seat of emotions, particularly fear) is larger.[19]

While neurologists are not certain what those abnormalities mean, one thing is clear: the brains of people with temporal lobe epilepsy are physically different from yours or mine. And the physical differences are probably dwarfed by the *functional* differences.

I thought of Jeffrey Saver's theory, that the repeated spiking in the temporal lobe's limbic system can place an exclamation mark after ordinary events, tagging them with cosmic meaning. For most of us,

August 15 is just August 15, but for Mary it is the Feast of the Assumption of the Blessed Virgin Mary. And at some point, this cosmic interpretation of events becomes the norm.

When you have occasional spikes from seizures, the UCLA neurologist explained, that extra electrical activity creates new connections between nerve cells. The more seizures, the stronger the connections. Eventually, the nerve cells form habits, like knee-jerk reactions of the brain. If your brain during a seizure tells you that certain events (or all events) have "God's will" stamped on them, then you will see God's will everywhere all the time, even when your brain is running normally.

This kind of brain, Saver told me, "is predisposed to have these types of experiences even between their seizures, because that brain had become more tailored toward religious experience."

Mary had no doubt her brain was attuned to religious experience.

"I know I'm very intuitive," she said, and I sensed she felt this was a divine gift, not the result of neural circuitry. "I have a friend who's a psychiatric social worker and she told me once, 'Mary, you can't enter into prayer with big groups of people. It's not safe for you.' She said, 'People like you are like very *loosely woven screens*. You need to protect yourself.' "

I looked up from my furious scribbling, reminded of Aldous Huxley's "reducing valve."

Mary smiled at me, then continued. "And I do. Because I pray for people and I start getting their symptoms. I prayed for a girl who's addicted to cocaine and other substances, and I woke up, and"—Mary pulled up her left sleeve, revealing marks and bruises covering the underside of her arm—"I had all these bruises."

She covered her arm quickly, then gazed at me.

"Mary," I said, a little flustered, "say I didn't believe in God. Say I thought this was just an electrical storm in your head. What would you say?"

Mary smiled kindly. "Well, I have a gift of faith. And I'm going to pray that you receive that gift, too."

A God Who Transmits

Many scientists would say that Jordan Sinclair and Mary are cursed with faulty wiring in their brains. If God is an electrician in this case, He miscalculated the voltage with tragic results. On a regular basis, the electricity in their temporal lobe surges and the circuit breaker flips on. The visions or transcendent feelings that the faulty wiring creates are nothing more than hallucinations.

But suppose the proper analogy is not an electrical storm but a radio transmission, in which the brain is a radio receiver. This thought did not originate with me. Several scientists I interviewed proposed the idea. In this analogy, everyone possesses the neural equipment to receive the radio program to varying degrees. Some have the volume turned low—in the case of an atheist, so low it's inaudible. Many hear their favorite programs every now and again. Others, through no fault of their own, have the volume turned up too high, or they are receiving a cacophony of noise that makes no sense, as if they were tuning in to stations transmitting from Atlanta and Montgomery, Alabama, while they drive through rural Georgia. If this analogy is carried further—and it is, by an increasing number of scientists—then the "sender" is separate from the "receiver." The content of the transmission does not originate in the brain, any more than the host of *All Things Considered* is physically in your radio. This is not to say that *all* our thoughts come from another, spiritual, realm, any more than everything we hear comes through the radio. It merely suggests that perhaps people suffering with an overactive temporal lobe are able to tune in to another dimension of reality, which the rest of us are unable to access. Maybe Saint Paul and Joan of Arc and Dostoyevsky were not crazy. Maybe they just had better antennae.

And maybe that was the case for Terrence Ayala as well.

Terrence smiled graciously when I met him at Henry Ford Hospital, even though I had kept him waiting for some time. A handsome forty-seven-year-old African-American, lean and muscular in his pullover sweater, Terrence spoke softly, almost shyly. I thought it a winning trait for a man with his résumé: Princeton University, the University of Virginia Law School, most recently a federal prosecutor in the Southern District of Florida.

Several years earlier, Terrence had undergone an operation that left him with a stuttering problem. His words, though they spilled drop by drop from his mouth, glistened with insight and originality. As I listened to him, I thought: *This is a man who sees the world differently.*

What Terrence sees is either hallucination or an alternate reality, depending on your point of view. Terrence told me that when he falls asleep at night, "frequently there's this dark presence, usually off to the upper right side of my body. If I'm lying down it will usually be looming over me. And I have a sense that it is a very evil presence. I can see it now—well, I can re-create that experience. It's a very palpable, powerful experience."

"And the presence seems just as real as, say, the walls, the table in the room . . . ?" I asked.

"Oh yes! It's as though there's another person in the room."

I remembered Michael Persinger's "Sensed Presence." We decided to call the presence "Bob." I asked Terrence if he thought Bob was ontologically real.

"Well, it's part of my reality," he responded.

"Do you think your brain picks up information that, say, mine can't perceive?" I asked. "I mean, do you think you can tune in to alternate realities, or states of consciousness, that I can't?"

"Absolutely," he said. He leaned forward. "If you look in a room full of people dancing, the majority of folks will be going to one beat, but there may be others who are tuning in to five or six or seven rhythms. And they can not only tune into them but move in consonance with those sounds, whereas other folks are just sticking to what others would agree is the dominant beat."

He laughed. "I wish I could tune in more frequently to the one everyone else is trying to dance to."

"When was the last time you saw Bob?" I asked.

"Not in the last two or three months," he said. "They changed my medication."

"Oh. They changed your medicine and that helped?" I asked.

"Helped?" he repeated.

I was startled. "Do you feel it's a loss?"

"I do," Terrence said. He had grown, if not fond of, then accustomed to his extra sense that he believed allowed him to tap into a different dimension.

"You know, we pay a lot of money to musicians who can hear different vibes and harmonies and express them for those of us who have not spent time training those abilities. We pay visual artists to make perceptible the images they have in their minds. And I guess this goes into a whole area of what we would call mental illness, and why we classify these things as *illnesses* rather than just *differences*. We have a habit of trying to bring people into conformity through medication and modern science and all kinds of things. *Who knows what realities we're medicating away?*"

This is the pivotal question. Are we medicating away realities or delusions? Science believes it has the dispositive answer. For, like magicians with their trick rabbits, scientists can now make these "realities" appear or disappear at will.

Recently a group of Swiss researchers was evaluating a twenty-two-year-old woman for possible brain surgery.[20] She had no psychiatric history. The researchers were homing in on a particular spot in the brain—the junction of the temporal lobes (emotional self) and the parietal lobes (the area that orients your body in space and in relation to other objects). When the surgeon electrically stimulated that area, the patient felt the presence of another person behind her. When they increased the voltage, she saw the "person" was young, of indeterminate sex, a "shadow" who did not speak or move. In the next stimulation, she observed a "man" sitting behind her, clasping her in his arms, which, she allowed, was rather unpleasant. Finally the researchers stimulated her brain while she performed a naming test, holding a card in her right hand. She reported that the man, now behind her to her right, was getting pushy (probably smarting from her earlier rebuff) and trying to interfere with her task. "He wants to take the card," she told the researchers. "He doesn't want me to read."

Stimulating alternate "realities" is a bit of a party trick. Making them disappear is far more common. Indeed, that is what epilepsy specialists are paid to do. It's called *treatment*. They lesion, or cut into, the brain and remove the offending tissue, or they medicate the brain and tamp down the electrical spikes. *Voilà*, the spiritual experiences disappear.

In fact, New York University neurologist Orrin Devinsky still remembers the day he made visions of Jesus go away. More than twenty years ago, when he worked as a resident at Memorial Sloan-Kettering Cancer Center, Devinsky was called to consult on a woman who had a brain lesion.

"She was talking about Christ, and Christ was speaking to her, and she was in a state of religious fervor, just talking nonstop about this," Devinsky recalls.

The brain scan showed an abnormality in her right temporal lobe, and her EEG showed that the region was experiencing continuous seizure.

"And when we used medications intravenously to shut down the seizure, the religious ideation stopped immediately," he says. "So here is a woman with a known structural problem—a lesion on her right temporal lobe—who had a well-documented seizure on her EEG, who had religious ideation that she could not control, it was just coming out of her in an almost violent form—not quite psychotic, but verging on that—and it was shut down immediately by treating her with medication."

You could conclude that if this woman's relationship with Christ vanished with a dose of medication, that experience was merely brain activity, not a connection with another reality.

But you could also argue that this proves nothing about the existence of God. Consider, once again, the brain as a radio. Let's say you open the lid and remove some components and wires: that's called *lesioning* the brain. Or say you pour maple syrup over the connections: that's called seizure medication. Either way, chances are the radio will not be picking up *All Things Considered*, even though the program is in fact broadcasting. In the same way, if there is an alternate reality, if there is a God who is constantly sending out signals, surgery or medication can destroy the ability to receive them—but God could still be speaking.

As I considered all I had learned about temporal lobe epilepsy, I realized I had circled back to the same irritating conundrum. Is the transcendent experience of an epileptic like music playing on a CD player—a closed loop dependent on nothing but the machinery? Or is it like a radio—tuning in to a hidden spiritual reality outside your physical brain?

Most scientists believe the question has been answered. It's all in your head. But that points more to the nature of scientific inquiry than to truth. Neurologists cannot climb inside Terrance Ayala's head and witness his alternate realities. Therefore under the rules of modern science—which require observation and precise measurements—Terrence's experience loses every time. It is a little like playing a basketball game in which the mystics' hoop is considered out of bounds. No matter how many times Terrence Ayala and Sophy Burnham and Don Eaton shoot the ball through the hoop, they cannot score points. But in fact, the ball still swished through the net.

Spiritual Experience for Bubba

Most of us have never suffered an epileptic seizure and resulting visions. But most have enjoyed some more run-of-the-mill moments of transcendence. I wondered: How do neurologists explain the experiences of "normal" people?

According to Michael Persinger at Laurentian University, people's spiritual experiences fall along a spectrum, with an atheist like Sigmund Freud sitting at one end and a mystic like Joan of Arc at the other. Somewhere in between are you, me, and anyone else who has shivered with a numinous experience. Persinger hypothesized that a person's spirituality correlates to his brain-wave activity. The more "temporal lobe signs" a "normal" person shows, the more likely he or she is to encounter God.

Persinger studied more than 1,000 people over ten years, comparing "normal populations" and "special normal populations" and clinical populations.[21] "Special normal populations" included creative people: artists, drama students, writers, and (oddly, I thought) women who experienced the psychological symptoms of false pregnancies. Clinical populations included people with epilepsy and post-traumatic stress disorder. He found a continuum of brain activity. The nonartistic drone showed average brain-wave activity, like the cold porridge of the mama bear. The artistic people showed more spike bursts: "theta" activity over the temporal lobes, which is a relaxed state of consciousness. They also presented discrepancies between the right side of the brain and the left. Persinger concluded that artists' elevated activity sparked creativity—the porridge was *just right*. But the clinical populations—they were too hot, and their brains' excessive temporal lobe activity brought anxiety, thoughts of suicide, and an intense, dominating, and often intrusive fantasy life.

What Persinger's work suggests is that anyone with a temporal lobe has a gateway to the divine, or at least to divine feelings. How strong those feelings are depends on the way your brain fires. You don't have to be crazy to feel "God." You just have to be human.

I asked Orrin Devinsky about my own mystical experience, a little baby experience, to be sure, but one that is seared into my memory and changed my view of the world. I told him about the incident eleven years earlier, when I had been interviewing a woman about her cancer and her religious faith.

"And suddenly I felt this numinous presence all around us," I recounted. "She felt it, too. Do you have any idea what would explain that?"

"I think the way you're wired, what was going on in your brain at that time, your past history—combined with that woman's story—profoundly affected you at some level," he speculated. "Maybe it kindled or reawakened a neural state that exists dormantly within your mind, and it triggered the right frequencies, and the right resonances, and the right connections, to reawaken that feeling. And the way that manifested itself within your nervous system was to trigger a specific type of religious experience."

"Does that exclude the possibility that there might have actually been something spiritual?" I asked. "I mean, does the fact that we might be able to explain it with brain function eliminate the possibility that there might actually be something spiritual going on?"

"No!" Devinsky said, and I was surprised by the heat of his answer. "I think the two can clearly exist together. Say there was a man and a woman who loved each other and when they looked at each other, they experienced that emotion that we refer to as love. There would be a change in their brain state, and probably a change in the temporal lobe as well. Does that negate the presence of true love between them? Of course not. When you get to spirituality, as a scientist I think it really becomes extremely difficult to say anything other than 'It's possible.'"

It makes sense that those epiphanies physically alter the brain as well as one's life. After all, being bitten by a dog, or memorizing 2 + 2 = 4, lays down permanent tracks in the brain. What might a moment with God do?

The brain of someone suffering with temporal lobe epilepsy is like a stallion that has not yet been broken and trained. It becomes too activated, it spooks too easily, and for those people, their spiritual experiences are wild, ragged, scary things, galloping through a forest with low-hanging limbs. For them, nothing but a shot of tranquilizer or an operation will calm down their wild lobes—and then, often but not always, the religious experiences disappear.

And yet I know people who connect with this "Other" and visit another reality without the drama of disease or the aid of drugs. These are the horse whisperers: they have trained their brains through years of meditation, and their restraining hand turns the wild stallion into an Olympic jumper or a Secretariat. These people are spiritual virtuosos, and none is more winning than Scott McDermott.

CHAPTER 8

Spiritual Virtuosos

NOT OFTEN DO MAGAZINE ARTICLES rob me of sleep. But the *Newsweek* cover story of May 7, 2001, gave me insomnia for a week. The article discussed "neurotheology," the neuroscience of spirituality—in this case, the brain activity of Buddhist and Catholic meditators when they dived deep into meditation. These experiments turned a corner in the study of spirituality. No longer would neurologists be reduced to merely extrapolating about spiritual experience from what they knew of abnormal conditions like epilepsy and psychedelic drug trips. They had moved into the realm of clinical, replicable research on spiritual events. And their subjects would not be mere mortals but spiritual Olympians, the Michael Jordans of meditation. I was hankering to get my hands into this research. It would take me another six years, but in June 2007, I would watch as a scanner took snapshots of my friend's brain while he communed with God.

I met Scott McDermott on January 20, 2004. We were sitting in a quiet room at the Toronto Airport Christian Fellowship Church. We could hear the quiet *brrrr* of hundreds of people speaking in tongues in the sanctuary below us. I was there on assignment for NPR to understand this boisterous brand of mysticism. Scott was ready to explain it to me.

Ten years earlier, the story went, the Holy Spirit had "fallen" on the worshippers during a Sunday-evening service. People began swooning and falling to the floor, where they remained pinned like upturned beetles for hours at a time. They began speaking in tongues, barking like dogs, and, most of all, laughing—laughing hysterically, hour after hour, sometimes for days. It would later be dubbed "Holy Laughter," and the phenomenon named the "Toronto Blessing."

I arrived for the tenth anniversary, hoping to witness the "blessing" in action. I was not disappointed. During the evening service, people in the audience began to chuckle, one or two at first; then the laughter rippled across the crowd like a squall across a lake. People fell at random, weeping piteously or laughing maniacally. I recorded people barking like dogs and clucking like chickens. I even heard a rooster or two. At the end of the service, people lined up to be blessed by a team of pastors. One pastor would touch a person's head, while another pastor stood behind to catch the worshipper as he toppled backward. They worked methodically, these anointed of the Lord, like woodsmen felling trees, one by one, row by row. I left well past midnight, with hundreds of people still sprawled on the floor, quietly clucking or barking.

At eleven the next morning, I sat down with Scott McDermott. A self-effacing man of slight build, his blond hair parted in the middle, Scott was the senior pastor of Washington Crossing United Methodist Church in Pennsylvania (about an hour north of Philadelphia). I expected him to provide an oasis of sanity and theological sophistication in this roiling sea of Pentecostal emotion. The United Methodists are known more for social liberalism than charismatic style, and Scott himself seemed hardly a "slain in the Spirit" sort of guy. He was a Ph.D. who in his spare time taught New Testament theology at Southern Methodist University.

But as he talked, his words flowed in a torrent of power and passion. I began to wonder if he was one of *them*, one of those people who fell on the floor. I asked him how he came to be at the Toronto Blessing. That is when he told me about his vision.

Scott first heard about the Toronto church in 1996, two years after the Holy Laughter had started there. He was skeptical.

"I'd seen enough of this kind of thing in American religion," he told me. "I did not want one more emotional experience. I was really put off. But enough people were asking me about it that I thought, Well, I should at least check it out."

Scott caught up with the Toronto pastor, John Arnott, at a church conference in New Orleans. At the end of the day, Arnott prayed for him.

"And when he did, I fell to the ground. And I felt waves of the Spirit flow through my body, from my hands to my feet, and back to my hands. Then the most unusual thing happened."

Scott paused, embarrassed, but the spigot had been opened and the story would not be stopped. He had a vision, he explained, that he was in Israel, overlooking the Judean desert.

"And I began to run on the floor. I'm on my back, and I began to run. I'm pumping my arms and legs, and I'm running from Jericho to Jerusalem. I did this for one and a half hours. And the other pastors in the room began to cheer me on. I saw them on the road in my vision. It was like the New York Marathon, people cheering, 'Run the race! Run the race! Run the race!' I ran as hard as I could past the Mount of Olives. I ran through the Garden of Gethsemane, through the Kidron Valley, and up the other side to what is the Eastern Gate of the Temple. It's presently walled shut, but there it was open. There was a finish line across it, and Jesus stood on the other side of the line with His arms outstretched, and I fell across the line into His arms and He just grabbed me and held me, laughing, holding on to me.

"At that point," Scott said, "I had stopped running on the floor, and

I felt the Lord say something to me that surprised me. He said, 'I want you to go with them, with my faithful servants.' But in my heart, I thought I didn't belong with them. For years, despite having a Ph.D. in the New Testament, despite teaching at a university, despite being a senior pastor of a pretty good-sized Methodist church, I didn't feel I measured up. And so I cried on the floor. I said, 'No, I do not belong with them. I do not.'

"And the Lord looked at me, with such intensity, and He said very firmly, 'You go with them.' And when He did, there was this indescribable love, it was beyond words, and it began to fill me. All that low self-esteem, all that sense of not being valued, began to melt in this love."

Scott paused to steady his voice.

"The change has been dramatic. I can't say I'm a perfect man as a result of this, or that every day I walk on cloud nine, but I will tell you this: Because of that, I know how much I'm loved by God. And when life gets hard—and life does get hard—God's love is there to see me through."

After his mystical marathon, Scott said he was horrified that he had made himself such a spectacle. But the next day, he said, "People would ask, 'Why would you have an experience like that and not me?' And I didn't have an answer to that one."

I think I do. Scott McDermott is a spiritual virtuoso.

Scanning the Mystical Moment

I thought about Scott McDermott often over the next three years, because he possessed an uncommon combination of qualities: the polished intelligence of a scholar and the all-consuming spirituality of a mystic. What was more, Scott practiced the presence of God, and his long, daily prayers seemed to give him access to another reality.

When I read in *Newsweek* that a scientist was studying the brains of

Buddhist monks and Catholic nuns, I thought of Scott. What was going on in Scott's brain when he connected with God, I wondered, when he saw the Kidron Valley or heard the voice of Jesus? I could not wait to get his brain scanned.

On June 1, 2007, Scott and I met in the radiology department at the Hospital of the University of Pennsylvania. We chatted for a few minutes, then followed a nurse down the hall to a cul-de-sac of examination rooms and one brain scanner. This was Andy Newberg's kingdom.

Andy Newberg is an associate professor in the department of radiology—with secondary appointments in psychiatry and religious studies—at the University of Pennsylvania. He is in his early forties but looks twenty-four. His curly dark hair has flecks of gray, but the rest of him cries out that the gray is an optical illusion. He gives the impression of a seventh-grader who went to bed thinking about his geometry test and awoke the next morning to find himself a neuroscientist.

Newberg has authored three books and countless articles on "neurotheology"—the study of the brain in the throes of spiritual experience. *Why God Won't Go Away*, which he wrote with his friend and mentor the late Eugene D'Aquili, explored the events in people's brains when they enjoy mystical experiences. This is the research that made the cover of *Newsweek*, the story that rendered me sleepless for a week.[1] Another Newberg book, *Why We Believe What We Believe*,[2] looked at different types of mystical experience, including speaking in tongues and the meditation of an atheist. He proposed a theory of why we seek out religious experience and how the experience cements our beliefs about God and the nature of reality. Newberg is a rock star in the small world of neurotheology, and certainly among the media, even though fellow neurologists sometimes tut-tut about the gleeful abandon with which he explores the human brain. Newberg doesn't seem to care. He's traveled way too far into the brain to turn back now.

For the past few years, Newberg has studied spiritual experts of all stripes: Tibetan Buddhist monks, Franciscan nuns, Sikhs, Pentecostals—

in other words, those who practice prayer and meditation long and hard. I had already witnessed him measuring the brain-wave activity and scanning the brain of a Sikh while he chanted his prayers. And on this first day of June 2007, I would watch as he scanned the brain of a Christian in prayer, my minister friend Scott McDermott.

One mystery about spiritually inclined people is this: What turns people into spiritual virtuosos? I have come to think that people like Scott McDermott are to prayer what Tiger Woods is to golf. From an early age, their natural talents channeled them toward golf, or toward God, and once they felt the rush of watching that ball fall into the hole or sipping the unearthly wine of transcendence, they pursued their passions. They trained because the reward was so sweet and so constant. Nature and nurture, genes and sweat, schemed to create these masters.

Scott was sixteen when he says he first "encountered" God directly. From that time on, prayer became the central habit of his life, around which all other events orbited. Scott has prayed one to two hours a day for more than thirty-five years. On Fridays, his day off, he is often on his knees for four. During those sessions, Scott feels peace and joy. He says he often hears a voice and receives visions.

"I begin early," he told me. "I begin at five-thirty. I'm not a morning person. I stumble down to my basement. My study's down in my basement. I have a recliner down there. I walk down, turn the lights on, and say, 'God, it's just me. I just want to spend some time with you.' And I sit in that chair and we begin to have a conversation."

I turned to Andy Newberg, who was standing with us in a small examination room at the hospital. Newberg was listening intently.

"I'm curious, Andy, what would that kind of practice do to the brain?" I asked.

"Well, some of the research we have been doing suggests that when people are engaged in practices over a long period of time, it does ultimately alter how the person's brain functions," Newberg responded. "As one does a particular practice or a particular task over and over, that becomes more and more written into the neural connections of the brain. So the more you focus on something, whether it's math or auto racing or football or God, the more that becomes your reality."

I glanced at Scott to see if he picked up the ambivalence in Newberg's response: God may be "your reality" and still be imaginary. The belief in God may shape your worldview, and your brain, in the same way Harry Potter books shape the brains and imaginations of children. But no one would argue that a wizard like Harry Potter actually exists.

I put the question to Newberg more explicitly: "When people pray, do they connect to God or tap into a dimension outside of their bodies?"

Newberg was prepared with a careful answer.

"Well, it comes down to belief systems," he said. "When a *religious person* looks at our brain scans, they say, 'Ah, that's where God has an interaction with me.' An *atheist* looks at the data and says, 'There it is. It's nothing more than what's in your brain.' Even if I do a brain scan of somebody who tells me that they've seen God, that scan only tells me what their brain was doing when they had that experience, and it doesn't tell me whether or not they actually did see God."

In fact, Newberg is undecided as to whether brain images reveal there is a God or not. Materialists would say that brain scans prove that prayer is a physical process, nothing more; there is no need to bring an external Being into the equation. But Newberg points out that brain scans do not necessarily exclude an external being. Say you are eating a piece of apple pie, just out of the oven, topped with melting vanilla ice cream. If Newberg took a brain scan of you as you bit into the pie, various parts of your brain would light up—the areas that register smell, taste, form, and shape, as would the area that recalls the memory of the time you tasted pie this good, at the county fair when you were six years old. The parts of the brain not involved with the task would go dark. Who knows? You may even lose a sense of self in this ecstatic culinary moment. But just because your brain is activated in a certain way, does that mean the apple pie isn't real? Of course it's real. And just because Scott McDermott's prayers correlate to brain activity, does that mean God is mere illusion?

Newberg then announced it was time for the study to begin. He closed the door and instructed Scott to relax on a metal bed in the sterile, cool room as a nurse inserted a catheter into his arm. They dimmed the lights. After ten minutes, Newberg tiptoed into the room and started the IV. A radioactive tracer began to flow into Scott's arm and up to his brain. After the tracer had set, Newberg led Scott to the brain imaging machine, which would take a snapshot of Scott's brain.

Scott lay down with his head in the machine. It was chilly in the room; the nurse tucked him in with a little blanket. For the next forty-five minutes, Scott rested with his head secured in place by pillows while the machine rotated around his head, taking images of his brain. The brain scanner then developed a color-coded photo of his brain: the active parts were hot red and screaming for attention; the ho-hum areas were yellow; the sleeping areas were a cool blue.

Scott went through the routine twice. The first time, Newberg told Scott *not* to pray but to allow his thoughts to wander, to remember a movie he saw, or chores he had to do. He could think about anything but God. This scan would produce a "baseline," or resting-state, image. In the second session, Scott prayed intensely for a particular person in what is called intercessory prayer. Like those weight-loss advertisements, Newberg would analyze the before and after, comparing the baseline image with the image of the brain in active prayer.[3]

After Scott emerged from the second brain scan, Newberg and Scott and I gathered in the examination room.

"Did you have any visual or auditory sensations?" I asked, thinking of his vision of running from Jericho to Jerusalem.

"Yes, I did have some of that."

"What sort of things?" I asked.

"I'm not sure you want this on tape," Scott said, laughing. It turned out he had been praying for *me*. He had a vision that God wanted to shatter the molds that I had put myself in (the wife mold, the reporter mold, and so on)—that God had a special mold for me.

"And I saw God handing you this mold of who you are. How He's made you. And He says, '*This* is how I want you to be.' It was intense, Barb. I would have screamed if I could have. It was that intense."

Newberg nodded thoughtfully. "We did a study recently of people speaking in tongues, and the whole concept is hearing what God has to say, and feeling the Spirit of God going through them." That sensation was caught on the brain scans.

"So, given your research," I asked, "what are some of the things you might expect to see on Scott's scan?"

Newberg was uncertain. Scott's prayer life seemed to straddle the two types of states that Newberg had studied. One type involved meditative practice, a state of mind a little like a champion dressage horse under perfect control. The other involved ecstatic Pentecostal prayer, which bucked and snorted like a rodeo bull. Which sort of brain Scott possessed would remain a mystery for several weeks.

Looking Under the Hood

So far, Newberg has identified two types of virtuoso brains: contemplative brains and rowdy ones. Let's examine the contemplatives first. For this study, Newberg put Franciscan nuns and Tibetan Buddhist monks (separately, of course) into the brain scanner at the University of Pennsylvania. The "theologies" underpinning these two practices have nothing in common. During their centering prayer—a meditative prayer that emphasizes interior silence—the nuns focused intently on God, and usually on Jesus. During their meditation, the monks nestled into a state of intense awareness, connecting with the underlying reality of life; their belief system excludes a supernatural, external "God." Yet each group's descriptions of those transcendent moments bore an uncanny resemblance.

"I felt communion, peace, openness to experience," Sister Celeste, a charming seventy-year-old Franciscan nun, recalled after emerging from the imaging machine.[4] She felt "an awareness and responsiveness to God's presence around me, and a feeling of centering, quieting, nothingness [as well as] moments of fullness of the presence of God. [God was] permeating my being."

Now listen to Michael Baine's words. Baine is a Tibetan Buddhist and a scientist who works with Andy Newberg and became one of the subjects in the study. Baine described his meditation experience as "timeless and infinite."

"There was an intense feeling of love," Baine told Newberg. "I felt a profound letting-go of the boundaries around me, and a connection with some kind of energy and state of being that had a quality of clarity, transparency, and joy. I felt a deep and profound sense of connection to everything, recognizing that there never was a true separation at all."

It sounded to me like two different road trips. One group drives a Lexus through the Redwood Forest, the other takes an Escalade through the Swiss Alps. The vehicles look entirely dissimilar, as does the scenery. But gazing at the giant trees and towering mountains may stir up a similar sense of exhilaration or awe. For Andy Newberg, the question was this: What is going on under the hood? Are the brain mechanics the same for Catholic nuns and Buddhist monks as they experience the transcendent, just as two different cars in two different countries operate along the same mechanical principles?

That is precisely what Newberg discovered when he peered at their brain scans. With both the monks and the nuns, the front part of their brains "lit up" as they focused on the task at hand. Think of the frontal lobes as the chief operating officer of the brain, one with accountant-like tendencies: it handles the details, helps plan and execute tasks, keeps you awake and alert and, above all, focused. It also processes memory and language and other complicated social tasks, such as figuring out how to behave at a cocktail party or how to respond to your spouse or friend or rival at work.

As the Christian nuns focused on God—on a word like *Jesus* or *Elohim* that helped them connect to the divine—their frontal lobes shifted into overdrive. Similarly, as the Buddhist monks meditated on an image that allowed them to connect with the ground of being, their scans showed their frontal lobes as a red glow of activity.

Newberg found another peculiar similarity. With both the nuns and the monks, the parietal lobes went dark during deep prayer and meditation. Newberg calls this the "orientation area" because it orients you in space and time: those lobes tell you where your body ends and the rest of the world begins. That is why Sister Celeste (and countless other mystics) described a unity with God, or as she put it, God "permeating my being." It was the neurological reason that Michael Baine felt "a deep and profound sense of connection to everything, recognizing that there never was a true separation at all." And, I might add, it was what those who enjoyed psychedelic drugs and natural mystical experiences reported.

Newberg theorizes that when the nuns and monks focused on their mantra or image, their brain simply screened out other information. You're watching *Casablanca* and the oven timer goes off, or you're gazing rapturously at your beloved and the phone rings—you don't notice. Increase that a hundredfold and you would lose your sense of time and space. It is not that the orientation area of the brain is not working. Rather, the frontal lobes are physically blocking all the information going to the orientation area—the sounds, the sights, the dog at the door or the timer in the kitchen, the things that would normally create a picture of the world around you.

And yet the orientation area, conscientious beaver that it is, is still trying to do its job. "It's still trying to create for you a sense of yourself and a spatial relation between you and the rest of the world," Newberg says, "but it has been deprived of the information that it normally has to do that, so you wind up with this sense of no self, no space, no time."[5]

Newberg's description reminded me of the way psychedelic drugs may behave in a brain to create hallucinations. Some pharmacologists believe that drugs like psilocybin block out external sensory information, allowing you to create your own, transcendent, reality. Chemicals are quicker, but it may be that prayer and meditation accomplish the same high—without the potential for a bad trip or ending up in handcuffs.

For me personally, Newberg's brain scans are theological dynamite. They boil down to this: a mystical state is a mystical state. The closer one draws to a transcendent state—or, as Newberg calls it, "absolute unitary being"—the more the descriptions merge. Christian mystics sound like Sufi mystics, who sound like Jewish mystics, who sound like Buddhists. And from the brain's point of view, this makes perfect sense.

"Buddhists and Hindus and Christians and Jews who have had mystical states tend to describe the states as 'everything becomes one.' The same terms keep cropping up over and over again," Newberg told me. "When we look at the physiology of the brains, the most unitary state is one in which we completely deprive the orientation parts of the brain of information. So, *physiologically* it should be very similar. And *philosophically* it should also be similar. If you have a totally undifferentiated experience, it's undifferentiated. It really has to be the same regardless of where you're coming from."

It doesn't matter if you scale the spiritual peak using Christian centering prayer, Buddhist meditation, or Sikh chanting. The destination is the same. Or as Sophy Burnham might tell you: You must choose a spoke to get to the center of the wheel, but any spoke will do.

Newberg's research throws a gauntlet at the foot of my faith. If disparate religions drive the same neural routes to transcendence, can one religion claim that it is true and all others false (or at least deficient)? I had noticed in my reporting that the people who experienced mystical states tended to drop religious labels: if they had been Christian before, they often became "spiritual but not religious" afterward, or they might incorporate other traditions into their practice of Christianity. One thing they often rejected, however, was an exclusive claim to Truth. This forced me to reconsider Jesus' statement, "I am the way, the truth, and the life. No man comes to the Father but by me." Perhaps, I reflected, Jesus' words were more nuanced than a literal reading of the text suggests.

Of course, I was not ready to throw my faith overboard just because of a few brain scans. Even if two things appear identical physiologically, they can play out in distinct ways. Identical twins share the same DNA, but they are not identical people. Can one routinely score 100 on math tests and the other flunk? Can one become a car salesman and the other an English teacher? Of course. Biology does not determine everything—and it certainly cannot determine the nature of "Truth." In the same way, it is too facile to say that brain states can determine the veracity of spiritual claims. And yet, I wondered, if there is a God who maps out the neural routes to communicate with Him, He could be making an explicit point: maybe the distinctions between religions are more artificial than believers want to admit.

Prayer Without Thinking

Andrew Newberg is an equal opportunity scientist. He is intrigued by brains in meditative states *and* in excited ones, which brings us to Pentecostal Christians. Anyone who has stood in a charismatic church and listened to the trilling of tongues could have told Newberg that these brains are nothing like those of contemplative nuns or Buddhist monks. The question is, how are they different?

What differentiates one type of spirituality from another is not doctrine—witness the scans of monks and nuns. What distinguishes one type of spirituality from another is style. And here we come to a very strange style indeed.

Since people started speaking unfamiliar languages at Los Angeles' Azusa Street Mission in 1905, Pentecostal Christianity has swept the world like a tornado, carving its way across the United States, turning south to Latin America, and most recently, carving a broad swath through Africa. It is the fastest-growing religious movement in the world. It knocks people down, "slaying them in the Spirit." It sparks laughter and apparent healing, and most of all, it confers upon people their own "prayer language." For the hallmark of a charismatic or Pentecostal Christian (the words are often used interchangeably) is glossolalia, or "speaking in tongues."

In his "terrifying" study, as Newberg put it, the scientist recruited five women who had spoken in tongues for several years. The protocol mirrored the one for Scott McDermott and the nuns and monks—with one major difference. Since Newberg wanted to detect changes specific to speaking in tongues, he asked each Pentecostal subject to sing gospel songs as the "baseline" state, and then to speak in her unknown prayer language as the "target" spiritual state. As in the other studies, a radioactive tracer was injected into the subject's bloodstream at the (presumed) height of her singing and glossolalia, capturing her brain in musical and then spiritual ecstasy.

Donna Morgan, one of the hospital's nuclear medicine technologists, told Newberg she had a deep interest in observing someone speak in tongues, and volunteered to help.[6] As they prepared the first subject, Newberg confessed to Morgan he was nervous, as he had never seen this phenomenon before. What if nothing happened? he asked her.

"Don't worry, it will," she reassured him.

Two minutes into the second session, the subject began uttering incomprehensible words, like a foreign language. She returned to English, then back to tongues. Newberg suddenly noticed Donna Morgan singing and moving around the room, and a few seconds later, she broke into her own prayer language.

"This is incredible," Newberg whispered to another assistant, and they stared in astonishment as the two women gaily babbled away for the next fifteen minutes. Morgan eventually became a subject and co-author of the journal article describing brain activity while a person speaks in tongues.[7]

The brain scans suggest why glossolalia is rarely heard at Harvard or Oxford. From a cognitive-processing standpoint, it is rather lowbrow. For when Newberg developed the brain scans and peered at the frontal lobes—the executive part of the brain, which manages the higher thought processes—he stared at the images in disbelief.

"The frontal lobes actually shut down," Newberg told me. "This actually made a lot of sense to us on reflection. When you're meditating, you kind of control the process. You focus on something, you attend to something, and you're willfully doing that. In speaking in tongues, the experience is that your will is not in charge of this whole process. These people have *surrendered*, and whatever happens just happens through them, there's nothing specific that they are in control of."

"Who is running the show, then?" I asked.

"Well, that's a good question," Newberg responded neutrally. "Obviously the spiritual answer is that it's the Spirit of God that is controlling this. From a physiological perspective, one might postulate that there's another part of the brain, a preconscious part of the brain that is causing these changes to occur. And that's why it *sounds* like language but it's not really language—because it is not tied into the cortical areas that would help you to produce something that is comprehensible. But we just don't know."

In other words, Saint Paul may have been describing a neurological reality when he wrote to the Romans, "We do not know what we ought to pray for, but the Spirit himself intercedes for us with groans that words cannot express."[8]

Newberg spotted another unexpected activity in the Pentecostal brain. In contrast to the brains of the nuns and monks, the activity in the parietal lobes (the association area) in the charismatic brains actually increased. While the nuns and monks lost their boundaries and merged into God or the universe, Pentecostals remained keenly aware of themselves as separate from God. It is a relationship, not a union, a finding that other "neurotheologians" have picked up on as well.[9]

In short, speaking in tongues is the physiological antithesis to Christian centering prayer. Despite their shared beliefs in Jesus as the Son of God, their spiritual practices have very little in common, both in the brain and outside of it—which is not to pronounce one right and the other wrong, but rather to suggest that there do appear to be many routes to transcendence.

A Spiritual Marker

Six weeks after Scott McDermott prayed in the brain scanner, Andy Newberg called me with the results. They were not dramatic, he said, but he found them a little surprising. Scott's frontal lobes decreased in activity, and the association area (the parietal lobes) increased. This meant Scott's brain behaved more like a Pentecostal than a contemplative nun.

I was not surprised. Scott had told us that when he prays, it is "dialogical," that is, a conversation. "When I'm praying for people, I'm just trying to hear God, and flow with God's heart toward that person. I don't feel a loss of myself at all."

In other words, Scott engages in conversation with Jesus, and does not "merge" into Him.

Moreover, when you hear Scott describe his prayer life, beneath the polish and the Ph.D. is a happy charismatic. Even when he was praying in a sterile examination room at the Hospital of the University of Pennsylvania, Scott told me he heard God's voice, saw a vision, and spoke to

God in tongues. So intense was the experience, he wanted to scream. In other words, it was a primal experience that did not require much heavy lifting in the frontal cortex.

However, what really caught Andy Newberg's attention was an inexplicable quirk he found during Scott McDermott's resting state. He spotted the same quirk during the resting state of every one of these spiritual virtuosos. It involved the thalamus—a tiny part of the brain that serves as traffic cop, taking in sights, sounds, and other sensory information (except for smell) and routing them to other parts of the brain. (You may recall the thalamus's role in bad mushroom trips: it fails to filter sights and sounds, allowing a nightmare of sensory overload to occur in the user's brain.) Newberg argues that the thalamus, armed with all that rich sensory information, makes the spiritual experiences feel lucid and *real.*

Here's the twist. In most people, the thalami (there are actually two, one on the right and one on the left) have the same level of activity. They beat along, side by side, like an old married couple. Newberg has analyzed thousands of brain scans during his day job at Penn, and they look pretty much the same. He scanned my brain, for example, and found only an anemic 3 percent asymmetry.

But in every spiritual virtuoso he studied—the nuns and monks, the Pentecostals and Scott McDermott—one side is more active than the other. Scott's thalami, for example, showed a 15 percent asymmetry. Newberg says that in his ten years of reading brain scans, he has never come across a similar finding. In fact, this kind of asymmetry was so rare that he searched for other such cases in the literature. The only similar cases he found were in people who had neurological damage caused by tumors or seizures.

"I think of it as a spiritual marker," Newberg told me. He confesses to be mystified about what purpose that asymmetry serves. Nor does he know whether people are born with a lopsided thalamus and the quirk somehow inclines them toward God—or whether their hours of prayer and meditation create the asymmetrical thalamus. But the finding does offer more evidence that spiritual brains are special.

It occurs to me that other "fingerprints" of God have been detected down the ages. Not inside the brain—for until recently we did not possess the technology to peer inside—but in the behavior and perception of those who claim to have touched God. One fingerprint was erotic: consider the ecstasy of Saint Teresa, who was reported to have orgasms when she prayed, or the sexual feelings described by Sophy Burnham. Another was sensory: Saint John of the Cross's sudden revelation that we are one with everything in the universe, and the same insight that swamped Arjun Patel as he meditated in his dorm room. A third was auditory: the voice heard by Joan of Arc, and the one that speaks when Scott McDermott prays. The fingerprints leave a lasting psychological mark, since the person undergoes a radical change in personality and ambition after he or she has touched the hem of God's garment. Usually, the change is lifelong.

History and theology have recorded these events, with little explanation except to say, here is mystery, or here is madness. But now, with our ability to peer into the brain, it seems the stories of people being felled and changed by the spiritual are the *outward* evidence of *internal* rewiring, just as unspoken ideas may take form in a painting or a book or a joke to a friend. I stumbled across one other neurological fingerprint of spiritual experience in my research, involving near-death experiences. To my mind this does not solve the mystery but uncovers another dimension, allowing us to excavate, like archaeologists who discover ancient civilizations, the hidden layers of spirit.

But wait—good news is at hand for the spiritual Luddites who are not graced with naturally mystical brains. If we are willing to pay the price of admission, we, too, can tune up our brains and go to spiritual destinations we never imagined. Why? Because our brains are plastic.

The Dalai Lama Meets the Neurologist

Geographically, the Dalai Lama lives in exile in Dharamsala, a remote refuge in India. Scientifically, His Holiness has a ubiquitous presence. In 2005, he drew a crowd of 5,000 scientists to hear him speak at the annual meeting of the Society of Neuroscience in Washington, D.C. Given his fascination with the nexus of science and meditation, it is hardly surprising that the Dalai Lama eventually heard of the work of Richard Davidson, a scientist at the University of Wisconsin who had studied the neural correlates of emotion. The collaboration of the holy man in the Himalayas and the tall gangly one in Madison, Wisconsin, would put neurotheology on the map.

Raised Jewish, Davidson had attended a yeshiva for seven years in Brooklyn before delving into Eastern philosophy as an undergraduate at New York University. In 1974, when he was a doctoral student in psychology at Harvard, Davidson ventured to India for his first meditative retreat. There he learned the mental rigors of Buddhist meditation and watched in awe as some contemplative monks sat hour after hour, sometimes fifteen hours a day, fully engaged in their internal mental world. As he did, a question arose that would chart the next thirty years of his life. Was there something about the monks' brains that allowed them to respond to "life's slings and arrows" more positively—and could anyone do the same?

"I became more and more interested in the possibility of transforming our brain by changing our mind, and in how meditation could play a very beneficial role in that process," he explained to me as we sat in his office, which looked out on the snowy midwestern campus in mid-February.

Davidson believed—and later demonstrated—that mental exercise could sculpt a person's mental circuitry, just as lifting weights could sculpt his biceps. Davidson had shown as much with the EEGs of Buddhist meditators, who could with a little focus shift their brain-wave activity to the left side of the brain. This intrigued Davidson, since earlier studies had shown that people with higher brain-wave activity in the *left* prefrontal cortex reported feeling more alert, energized, enthusiastic, and joyous. People with higher brain-wave activity in the *right* side reported feeling more worry, anxiety, and sadness; they rarely felt elation or joy. The fact that the brain-wave activity of the Buddhist monks *swamped* to the left persuaded Davidson that the brains of these meditators were different from yours and mine. The question thus became: Are these Buddhists born with different brains, which is why they gravitate toward meditation? Or could anyone achieve that state of joy, peace, and holiness with a little practice?

Enter the Dalai Lama. When the Dalai Lama heard of Davidson's work, he invited the neurologist to Dharamsala for a chat. So it happened that in 1992, Davidson, two other neuroscientists, and a Buddhist scholar dragged hundreds of pounds of equipment—laptop computers, EEG machines, and untold numbers of batteries—to a remote mountain refuge. Their mission: to measure brain-wave activity of Buddhist "adepts." These monks had between 10,000 and 50,000 hours of meditation under their belts. If mental Olympians did exist, they would be found there, under the wing of the Dalai Lama.

"From a scientific perspective, it was thoroughly unsuccessful," Davidson recalled. "Most did not speak any language other than Tibetan. They have not lived anywhere other than in the Himalayas. They had never seen a computer before. To many of them, just the act of interacting with a keyboard was completely novel. And so it was a rude awakening for us regarding the problems of bridging this cultural chasm."

The scientists left India with no data, but they did leave with a prize far more valuable in the long term: they piqued the interest of the Dalai Lama, who eventually sent eight of his monks to Davidson's laboratory in Madison, Wisconsin.

They arrived one by one, dressed in their saffron robes and eyes wide as pies, to be slid into brain scanners, or affixed with 256 EEG electrodes on their shaved heads, hanging down like dreadlocks. Davidson would compare the brains of these monks with the brains of ten students who received one week of meditation training. In the study, the subjects were shown emotionally evocative photographs, such as a baby with a horrendous tumor on its eye, or a wailing man carrying his son away from an explosion. They were told to focus their minds on unconditional compassion, and a readiness to help all living beings.

Davidson could have predicted the results, but it was nice to have hard evidence: to wit, the human brain can be molded and changed. It is *plastic*. The students' brains shifted slightly between the times they were resting and the times they were engaging in compassionate meditation. But when the *monks* viewed the photographs, the parts of the brain associated with empathy and mother's love lit up like Times Square. Ditto for the area of the brain involving planned movement: their brains were saying, *Hey! Get up—do something!* Moreover, their brains confirmed what the monks already knew. They were happy monks: the left prefrontal area was a cauldron of activity, to a degree never seen from pure mental activity.

More intriguing, to me at least, was a certain kind of brain rhythm called the gamma rhythm, which is extremely fast and associated with alertness and attention. The monks' brains were flooded with gamma waves, not in one area, but all over, and this synchronized or knit together disparate brain circuits. This produced a rare state: one writer describes it as an "ah-ha" moment[10]—when your brain brings together the sound, the look, the feel, the memory of an object, and then . . . Aha! (*That voice, that face—Oh! That's Hugh Grant on the television,* or, *The smell, the color—Oh! There are burgers on the grill. I'm hungry.*) While ordinary mortals enjoy that moment of recognition (synchrony) for a few milliseconds, Davidson said these monks were able to sustain it for more than *five minutes*.

In other words, Davidson, like Newberg, found neurological fingerprints in his monks when they were meditating. And like Newberg, he also found a *permanent* neurological fingerprint among his spiritual virtuosos. When the monks were resting, their brains still resided in the hyperalert, synchronized, happy zone—just less intensely, as if the volume had been turned down. This strongly suggests that meditation had permanently altered their brains. In essence, Davidson's gamma rhythm may be a fingerprint of the meditative experience, an indelible mark that something strange and wonderful has happened.

Little League Meditators

Davidson's research indicated that well-trained "spiritual" brains operate differently from run-of-the-mill brains, but he also suspected that with enough exercise, any normal brain can scale unimagined spiritual and neurological heights. Fine. No doubt many people would like to climb Mount Everest, but who has a year to spend in preparation, acclimating to the altitude and developing the calf muscles to carry a hundred-pound pack? Who can afford to devote 10,000 hours to meditation just to alter the brain circuitry in their head?

But even as he put meditative Olympians through their paces, Davidson and others were looking toward mere mortals with jobs and kids. They suspected that with only a little training, ordinary people could also remold their brains and their outlook on life.

The employees at Promega Corporation, a biotechnology company outside of Madison, worked long hours under stressful conditions. They were typical people working in the high-tech world, and perfect subjects to test the hypothesis that a little mindfulness can change your life. And that is how twenty-five employees at Promega—scientists, laboratory technicians, marketers and managers—learned to meditate.

Every week, Jon Kabat-Zinn traveled from Boston to Madison,

carrying a boom box, a set of meditation tapes, and his Tibetan chimes.[11] Kabat-Zinn pioneered mindfulness training from his post at University of Massachusetts Medical School. Once a week, for two and a half hours, the employees sat on the floor, practicing "mindfulness." They were also expected to practice the meditation technique on their own for forty-five minutes a day. They were taught how to look at their thoughts and emotions clinically, like an outsider, and test their external reality. Is this project really killing me? Will winning that account solve all the problems in my life? Mindfulness, Kabat-Zinn told them, could free them from the bondage of their emotions. Sixteen other employees would serve as a "control group," receiving no training until after the study was completed.

Kabat-Zinn, Davidson, and other researchers wired up all forty-one people to EEGs to determine their brain-wave activity. Before the mindfulness meditation, the employees' "set point"—or natural attitude—was anxious and worried. That was reflected in their brain activity, which tipped to the right prefrontal cortex, the anxious, worried section of the brain. After eight weeks, however, employees steeped in meditation said their moods had improved; they were feeling less anxious and more engaged at work. When they were hooked up to the EEG again, their brain-wave activity had shifted leftward, to the "happy" part of the brain. In little more than two months, they had traveled from the bleak grays of February to the vibrant colors of May—a seasonal shift that took place not in the weather but in their minds. The control subjects remained in the anxious zone the entire time.

The employees of Promega had rewired their brains and moved their set points, for the affordable price of forty-five minutes a day. Over the past couple of decades, scientists have been persuaded that the brain is "plastic" and moldable, even into late life. But for something as "fixed" as a set point to be neurologically unstuck and moved to a happier zone in *two months*—well, who knew that changing one's mind could alter one's brain so quickly?[12]

Now we are talking about a reasonable time frame, I thought. And when I learned that one of Davidson's research assistants named Helen Weng was trying to determine if the meditation could change the brain in as little as *two weeks,* I wanted in. In journalism, where a news story has a shelf life of a few minutes, two weeks is a frivolously long time to concentrate on any one subject. I could handle two weeks.

So I signed up. Or I tried to. It turns out that at forty-seven, I was too old to be included in the study. They had set an arbitrary age limit of forty-five. Still, I decided to practice the compassion meditation training every day for two weeks, and see what happened.

For thirty minutes every morning, with the house dark and still, I put on my headphones and absorbed the soothing voice of Helen Weng. *May you be free from suffering,* she intoned gently in a taped message, *may you have joy and peace.* In the first part of the exercise, I was instructed to wish happiness and freedom from suffering on a loved one, noticing any physical sensations around my heart as I did so. Next, I was to shower myself with compassion; then repeat the exercise for a stranger, someone I did not know but saw on occasion; and finally turn my compassionate thoughts toward a "difficult" person. At the end of each session, I wrote down my thoughts and feelings.

I was a poster child for meditative failure. I excelled at wishing well of my loved ones. I recalled wrenching episodes involving my mother or brother or close friend, even cried a few times as I lived their suffering. I enjoyed showering compassion on a stranger, making up elaborate stories about the young lost soul at Starbucks, the mailman, the colleague at work who was diagnosed with cancer. The problem arose during the rest of the session. When I recounted sad events in my life and tried to relieve my own suffering, sometimes I began to sniffle; then I grew ticked off at the person who caused the suffering. By the time I had meditated for thirty minutes, I was feeling sorry for myself, in a foul mood, pouring angry screeds into my journal about the injustice of my life. My husband kept a wide berth during those two weeks. He was visibly relieved when my compassion training ended.

"So, is there a capacity to change my brain if I continue doing this?" I asked Richard Davidson when I met him a few days later. I did not disclose how ill-tempered I had become.

"Absolutely," he assured me. "I would say the likelihood is that you are already changing your brain, be it very modestly, in ways we may or may not be able to measure."

I shuddered inwardly.

Davidson smiled. "Keep practicing," he said.

The Mind and the Brain

In the scores of interviews for this book, I noticed a predictable chasm between people who had experienced transcendence and those who had not. Both would burn at the stake for their positions.

On the one side marches the well-armed, highly trained, battle-tested brigade of scientists who insist that everything is caused by material processes. These scientists—and they are the vast majority of the academic community—believe that thoughts, feelings, desires, and intentions arise from the interaction of brain chemicals and electricity. They arrive at this conclusion through observation, using material instruments to measure a material brain. They echo the assertion of Nobel Prize winner Francis Crick, who discovered the structure of DNA with James Watson. In his book *The Astonishing Hypothesis*[13] he stated, "You, your joys and your sorrows, your memories and your ambitions, your sense of personal identity and free will, are in fact no more than the behavior of a vast assembly of nerve cells and their associated molecules."

On the other side of the debate is a small, underfed, and under-armed guerrilla force lobbing single grenades from the bushes. These scientists insist that cells and molecules do not determine all of human existence. They claim that molecules do not explain love, or willpower, or the occasional glimpse into spiritual dimensions. Significantly, I noticed that scientists who had *themselves* waded into spiritual waters—through prayer, meditation, or a near-death experience—*always* fall into the spiritual camp. Their personal experience trumps the assumptions of modern science.

That is what made Richard Davidson such an enigma. Despite being steeped in meditative practice for thirty years, despite his close friendship with the Dalai Lama, he remains certain that everything boils down to material stuff. Meditation trains the mind to appreciate "the interconnectedness, the sense of there being a larger purpose," he told me. But in the end, he said, it is nothing more than brain activity.

Okay, now I was really confused. *Interconnectedness* and *larger purpose* sound like pretty metaphysical concepts to me. Forget about the "spiritual" realm, I said. Let's ratchet it down a level: Can one's thoughts—mind states—affect one's brain states? Can thought affect matter? Wasn't that the take-away from the studies on Buddhist meditators and even the biotech employees—that their meditative thoughts affected their material brains?

Apparently not, according to Davidson. "When you say mind states affect brain states, what we're really talking about is how certain parameters of the brain can affect other parameters of the brain," he said coolly. "When we engage in the process of training our mind, what we're engaging in is the process of using our *brain* to change our *brain*."

A few days later I called Davidson and asked him to elaborate. I could not quite grasp how the three-pound mass called my brain dictated everything I felt, thought, or did.

"Let me give you a simple example," I said. "I wanted to call you to ask you some follow-up questions. Can you explain how my *brain state* caused me to pick up the phone and dial your number? Where did that intention, the desire, arise from in the first place?"

"It's fully explainable, based upon the prior conditions and circumstances to which an individual is exposed." Davidson sighed, for surely I was the dimmest of students. "In the case of the phone call, you have reminders that you're supposed to make this phone call, you have a calendar, you see these cues, those cues elicit the intention. There's nothing magical about it."

Afterward, I kept thinking, I'm just not smart enough to understand that we are wholly material organisms driven helplessly by material interactions in the brain. I know it will all become clear eventually. All the while, my intuition rebelled. This reductionist thesis seems to contradict life as we experience it every day. It ignores free will and choice, the elements that differentiate you from your parakeet, the fact that I chose to marry Devin and not Lee, to force myself to go running in the rain, to spend my vacation revising this book, which certainly had no immediate evolutionary purpose. How could blind molecules in my brain decide those issues?

I put the question to Matthieu Ricard, a scientist who has coauthored some of Davidson's studies. Ricard is a Buddhist who has meditated for, on average, two hours and twenty minutes a day, seven days a week, for the past thirty-five years. He lives in Dharamsala with the Dalai Lama and serves as his translator. Ricard told me that the idea that brain and mind are the same—that we are a bag of molecules and our brains merely react to stimuli—makes no sense to him. Consider free will.

"If you say, 'Okay, just to prove my free will exists, I'll do something completely against nature and biology. I'm going to sit on my chair until I collapse. I'm not going to go to the bathroom. I'm not going to eat. I'm not going to drink.' A functional brain—except for a crazy person—would refuse to do that. So where does that free will come from? It doesn't make any sense from the body's perspective. The brain has no reason to say, 'I'm going to sit there and pee in my pants and not eat anything'—it's totally meaningless, except to prove that I am in charge."

Listening to Matthieu Ricard, I had a naughty thought. *Maybe the materialists have got it wrong.* Maybe the emperor has no clothes. Maybe we do have a mind, consciousness—a soul—that works with the material brain but is independent of it.

The attempt to explain spiritual experience through neurology alone reminds me of a joke I heard recently from a Buddhist monk. A person loses his car keys. It's dark outside, and he's looking right under a streetlamp. Another person comes over and asks, "What are you doing?" And the man replies, "Well, I lost my keys." And the other fellow says, "Did you lose them here?" And the man says, "No, I lost them way over yonder, but this is where the light is."

Brain activity, chemical reactions, the functions of the various lobes of the brain—this is where the light is for modern-day scientists. Peering at brain scans and EEGs is something they are really good at. And so they keep on doing it, even though there is a possibility that at least part of the explanation lies somewhere else, just beyond their circle of light.

Since scientists have not presented an airtight case that matter is the sum total of reality, I decided to venture beyond the circle of light, beyond the boundaries of safe, mainstream science. This leads directly to the ground zero of scientific debate: the nature of consciousness. Can a person perceive when her brain is not *functioning*? When the radio receiver is broken, or out of batteries, can it pick up a signal? What happens to consciousness when the physical brain is stilled by death?

CHAPTER 9

Out of My Body or out of My Mind?

THE DILEMMA FOR CONSCIOUSNESS RESEARCHERS IS THIS: How do you tease apart the mind from the brain? For most of us, the brain and the mind cannot be distinguished: if the brain stops, so does the person's consciousness and perception of reality. Little surprise, then, that most scientists believe the physical brain and nonphysical consciousness are two sides of the same coin.

But a healthy number of people say that assumption is flat wrong. No doubt you have met a few of them, perhaps unwittingly: people whose consciousness kept ticking after their brains shuddered to a halt. These people have approached the border of death and lived to tell the tale, and their stories suggest that mind is more than matter.

By his own account, Michael Sabom was an unlikely candidate to launch a methodical investigation into the murky world of consciousness. In 1976, a friend gave Sabom a copy of Raymond Moody's blockbuster book about near-death experiences, *Life After Life.* Sabom, a cardiologist, announced his assessment.

"I said I thought it was hogwash," Sabom told me.

But his friends challenged him to ask some of his patients whether they had experienced this phenomenon or not.

"I fully expected that they would not report anything like this. I went in and the third patient I talked to had had a classic near-death experience."

As he related his story, Sabom was sitting in a big brown leather chair in his brilliantly lit, uncluttered office in north Atlanta. His thick gray hair was as unruly as a Brillo pad, and his voice was southern and drowsy. He wore a white lab coat, a reminder that his day job was fixing people's hearts, not tracking their spiritual experiences. But he had cleared two hours in his afternoon for me. For more than three decades, Sabom has enjoyed a passionate affair with near-death experiences.

As a scientist, Sabom sought some proof of their existence. He realized you cannot verify the drama in someone's brain during a near-death experience, any more than I can confirm the veracity of the daydream you claimed to have enjoyed yesterday afternoon. Even if you had been lying in a brain scanner at the time of your alleged daydream, I still could not verify it, since we have not developed the technology to peek into your thoughts.

No, Sabom needed *external* evidence, something that could be corroborated. He settled on testing out-of-body experiences. Those are the moments before the famous "tunnel" and the white light and the conversations with dead relatives, when the heart has stopped, the patient is comatose, and, as he later claims, he leaves his body and hovers near the ceiling to watch the chaos below as doctors try to resuscitate him. If the patient's visual description of those events matched what actually occurred on the table, Sabom figured, that would go some distance to proving that a person's consciousness and identity do not depend on three pounds of tissue called the brain. It might throw a very heavy wrench into the materialists' position.

Between 1976 and 1981, Sabom conducted what is still considered the most meticulous research on near-death and out-of-body experiences.[1] At the time he was a cardiologist at the University of Florida Health Science Center in Gainesville, and he interviewed every one of his patients who had suffered a cardiac arrest. His sample included others who heard about his study. Of the 116 patients who reported near-death experiences, thirty-two claimed to recall details of their resuscitation during an out-of-body experience. Sabom queried them about their memories, and because he had access to their records, he could check their descriptions against the reports of what actually occurred when they were revived.

Some patients gave descriptions too generic to be of value. But other patients remembered striking details. One man, a security guard, described his collapse in the hospital hallway, the doctors' attempts to defibrillate him, how they plunged a needle into his heart "like an Aztec Indian ritual," how they tried to start an IV on his left wrist, but, realizing that hand was broken, switched to the right. The records confirmed this scenario.

Another man described his experience from a few years earlier, when he was caught in a firefight with the Viet Cong. He said he watched from above his body as the enemy soldiers left him for dead, surveying the carnage of his lifeless buddies. He described the Army medics placing him in a body bag, transferring him to the truck and depositing him in the morgue. He watched as they cut off his bloody shorts and made an incision in his left groin, to inject the embalming fluid into his femoral vein. He recalled the relief he felt when the medic noticed he had a pulse and resuscitated him.

"At the end of the interview I said, 'Do you mind if I examine your left groin?'" Sabom recalled. "And so he said, 'Sure, that's fine.' And sure enough, there was a scar right there where they had cut to go into the left femoral vein to go in to embalm him. So that was evidence to me that what he was telling he was not just making up." Sabom also obtained the medical records, which confirmed this and other details.

A third man, a forty-four-year-old retired Air Force pilot, was transfixed not only by his own resuscitation—he saw them pounding on his chest, breaking a rib, placing a green oxygen mask on him—but also by the machinery. In May 1978, he suffered a massive heart attack. He described the dials on the defibrillator, how the doctor called out the "watt seconds" as a "fixed" needle determined the voltage the doctor wanted, and a "moving" needle told them when the machine was properly charged. The man claimed never to have heard these terms before, nor ever to have seen the procedure on television.

"Couldn't he have heard the instructions and made a model in his mind?" I asked Sabom.

Not in this case, Sabom said, because auditory clues would not have been enough. "Somebody is not standing there saying, 'Okay, now watch this one needle as it goes up . . . stops . . . okay now . . .' That's not being discussed, it's just happening. So you either see it, or you don't know about it, because it's not verbal information that's being discussed at the time."

When conducting his research, Sabom did wonder if the patients might have made educated guesses—based either on their own experience with cardiac resuscitation or on television programs (although it would be a decade before realistic medical dramas took to the airwaves). To check that hypothesis, Sabom conducted a mini-study. He had in hand the interviews with the thirty-two patients who claimed they "watched" their resuscitation from outside their bodies. He then asked twenty-five "control" patients—individuals who had been resuscitated but remained oblivious throughout the procedure—to imagine being revived and tell him what that would look like.

"And twenty-three of the twenty-five [control subjects] made major mistakes in what they were telling me," he reported.

They muffed details about how the paddles were used, the sequence of steps during the resuscitation, where needles were inserted, how nurses drew blood gases from the wrists. Then he compared those accounts with the descriptions by the people who had claimed to have watched their resuscitation.

"There was just no comparison at all," he recalled. All in all, the group that claimed to have had out-of-body experiences was spot on. "Again, this is just some evidence to suggest what these people were telling me was coming from a different source of information than from something they knew about already."[2]

Targets out of Range

A skeptical scientist would likely say that no matter how it is accessorized with medical reports and tables, Sabom's research is only a pile of retrospective anecdotes. Therein lies the dilemma: How does a researcher *prove* that consciousness can gallop free, even when the brain is crippled or dead?

It took nearly fifteen years for researchers to arrive at a plan. It needed to be dirt cheap; after all, who would fund this research? And it should have a shot, at least, of building an airtight case that someone had left his body and perceived an object or event that he simply could not have seen otherwise.

I think of this plan as the "squirrel monkey test." When I was about eight years old, my family owned a squirrel monkey, imaginatively named Monk. One morning, Monk disappeared, and an entire day's search ended with the primate still missing. In the evening, my mother hatched an idea: Place a grape on the floor of every room, close the doors, and wait for a grape to disappear. Within minutes, the grape in the basement was gone. My brother then replaced the grape and hid behind a chair with the cage. The denouement occurred when Monk emerged to fetch that "target" grape, and the rest is, well, a blur of a leaping boy and a cage snapping shut and a startled monkey clinging to a grape that cost him his freedom.

In similar fashion, near-death-experience researchers decided to place a visual "target" in rooms (such as operating rooms) where people were likely to suffer cardiac arrests. Usually these were hung or placed below the ceiling but above the operating table—somewhere in the line of vision of an out-of-body-experiencer "hovering" near the ceiling. In one study, the target was a large laminated piece of cardboard with a pattern or a word on it that was changed every few days. In another, the researcher hung computer screens that displayed images of words or colors or pastoral scenes, rotating as in a photo gallery. The nurses and doctors who crawled up on a ladder to affix these targets hoped that a traumatized patient in the operating room would report floating above his body as the doctors worked on him, see the target, and then volunteer the information later. Would that not serve as irrefutable proof that the out-of-body experience had occurred? Wouldn't that prove the monkey was in the basement?

"It was a lot of work," Penny Sartori told me. Sartori, an intensive-care nurse, religiously rotated targets in the intensive-care unit at Morriston Hospital in Swansea (Wales) for five years.

"I had to decide how to do it," the diminutive thirty-five-year-old said in a lilting cadence. "I had to decide on which signals I was going to use. I had to go to the trouble of getting them laminated. And then when I put them on top of the monitors, every month I'd have to clean them for infection-control purposes, and rotate each to a different bed area. It was a lot of work. And no one actually saw the symbols. So I was very disappointed, yeah."

Out of five studies conducted in Europe and the United States, not one of the patients spotted the target.

Why, I asked Sartori, did she suppose, no patients spotted them?

"Well, a lot of the patients didn't float high enough to see the symbol," she said earnestly. "Some of them floated in directions opposite of where [the targets] were situated. And the patients who were high enough said they were so concerned with what was going on around their immediate body that they weren't looking anywhere else. I'm sure I'd be the same," she added, then smiled. "But one man did say to me, if I had known there would be a symbol there, I'd have gone up to it and looked at it and I would have come back and told you what it was."

A comforting sentiment, that, but no monkey. Not yet, at least.

The White Crow

I realized before I wrote the first word of this book that I would never be able to "prove" that God exists, or that the soul survives death, or even that the universe is an intelligent, caring place. One arrives at those conclusions through personal experience, through an encounter with a dimension of reality that just does not fit Newtonian physics. But as I delved further into the research, I picked up the scent of a provable story: a case that demonstrates that one's mind can be untethered from the body, and consciousness can fly free of the brain.

Harvard psychologist William James once said, "If you wish to upset the law that all crows are black, you mustn't seek to show that no crows are; it is enough to prove one single crow to be white."[3]

I found my white crow.

On the day I visited Pam Reynolds, the sky sparkled porcelain blue, the last breath of gentle weather before winter took its hold. It was October 30, 2006. I turned onto the gravel road to find a large brown and tan touring bus parked in the driveway. This was Pam's home. Pam is a musician, owner of Southern Tracks Recording in Atlanta, which has recorded the music of Bruce Springsteen, Pearl Jam, R.E.M., and Matchbox 20, among dozens of other musicians. Her son greeted me at my car cheerfully and opened the pneumatic door to the bus. I climbed the stairs and peered through the dim, smoky room, spotting a kitchenette with the remains of breakfast, a small table, couches lining one side of the bus, with a large-screen TV at the front.

"Hello, hello!" Pam called, a beacon of life in the dark interior. I made my way to the back of the bus. Michael Sabom had discovered Pam's case and written it up in his book *Light and Death*, but I had no idea what to expect. There, sitting cross-legged on a double bed that filled the back room, sprawled a redhead with shoulder-length hair and a wry smile. The sheets were mussed, as if she had just risen, and she was gamely attempting to tie her sneakers.

She gave up on her shoes, and I noticed she was struggling to breathe as well.

"Are you okay?" I asked.

"Oh, yes, baby, I'm just fine. A little vertigo. You're spinning in circles right now. And the adrenaline makes it hard to breathe. It's a fight-or-flight thing—vertigo telling you you're gonna fall, and the adrenaline kicks in, makes my heart beat too fast. It's worse to watch it than experience it."

Pam relaxed happily on the bed, not eager to move. She was fifty and looked far too young to have five children and four grandchildren. Her hair had turned gray after her near-death experience, she confessed, and now she dyes it reddish brown.

I asked if she ever goes on the road anymore, performing. She said she travels but can't perform. She gets vertigo, faints easily, walks with a cane. Then we settled at the table by the kitchenette with a cup of tea and she told me about the time she died and came back.

"At twenty-five I was a singer, songwriter, did some production, classical composition," she began. "I was busily being a mother and doing the suburban thing and working, and I began to experience excruciating headaches."

They grew worse each year and medication brought no relief. In the summer of 1991, when Pam was thirty-five, she and her husband, Butch, were promoting a new record in Virginia Beach, "and I inexplicably forgot how to talk. I've got a big mouth and I never forget how to talk. I forgot how to talk."

Pam and Butch rushed back to Atlanta. Her neurologist found a basilar artery aneurysm that was smack in the middle of her brain stem, the area that controls basic life functions, such as breathing and swallowing. And the wall of the aneurysm—like a bulge on a tire—was thinning. It was already leaking blood into her brain. As Pam put it, "there was a bomb in my brain that had already begun to explode."

The doctor suggested she get her affairs in order. But Pam's mother happened to hear of a "brilliant young man" in Phoenix who had pioneered a remarkable new procedure, and gave him a call. Neurosurgeon Robert Spetzler urged Pam to fly out to Arizona. He would perform the surgery for free. Two days later, Pam arrived at the Barrow Neurological Institute early in the morning.[4] By seven-twenty on that August morning in 1991, a team of doctors had wheeled her into the operating room and the anesthesiologist was administering a cocktail of drugs.

Pam then began a surgically driven journey to the edge of life and back, called a "standstill operation." In the next four hours, Dr. Spetzler, assisted by twenty doctors and nurses, taped Pam's eyes shut and placed cooling blankets around her body, packing her with ice to put her in a deep freeze. As her body temperature began to plunge, the cardiac surgeon inserted a Swan-Ganz catheter—like a long piece of spaghetti—into her jugular vein and threaded it to her heart, then attached Pam to a heart-lung machine.

When her body dropped to around 80 degrees, Pam's heart began to falter, at which point the doctors administered massive doses of potassium chloride. This stopped her heart completely, and left her wholly dependent on the machine. Pam's body temperature continued to plummet.

"As the temperature gets colder and colder," Dr. Spetzler told me in an interview, "we get to a point—usually around 60 degrees—where we can turn off the machine, and actually drain blood out of the body."

They drained all the blood from Pam's head into "reservoir cylinders," similar to draining oil from a car. The aneurysm sac collapsed for lack of blood. "We can then expose the aneurysm and clip it."

"At this point," I asked, "could Pam see or could she hear? Could you describe her state?"

"She is as deeply comatose as you can possibly be and still be alive," Spetzler replied. "Now, how do we define that? First is the anesthesia that puts her to sleep. Then we give her medication—barbiturates—which knocks out her deepest brain functions. And how do we know that? Because we monitor an EEG and we monitor evoked potentials," which measure brain-stem activity.

The device used to measure brain-stem activity was a set of molded ear speakers, affixed with mounds of tape over her ears. These emit loud clicks of 90 to 100 decibels, equal to the sound of a jet plane taking off.

"As the brain goes deeper and deeper into sleep, it becomes less and less of a signal, and in her case, they [the vital signs on the monitors] go completely flat," the neurosurgeon explained. "So not only is she given medication to put her into the deepest coma, but then you add on this hypothermia, which puts her into an even deeper coma. Her brain is as asleep, as comatose, as unresponsive, as it can possibly be."

"She wasn't technically dead, though, was she?" I asked, anticipating the criticism of skeptics.

"It's an artificial definition," Dr. Spetzler explained. "If she were awake, and she had no pulse, no blood pressure, no respiration, we would call her dead. But if you are in this suspended state, because we know you can come back, I would not define it as dead."

"During this time, could she see anything or hear anything?"

"Absolutely not."

After Pam's aneurysm was clipped and removed, the doctors reversed the process, warmed the blood, introduced it into her body, and at around 78 degrees, Pam's heart began beating on its own.

When she awakened, Pam had a story to tell.

The View from Above

After the doctors administered the anesthesia, Pam told me, "I barely remember going to sleep. And I was lying there on the gurney, minding my own business, seriously unconscious. Dr. Spetzler said I was in a deep coma—when the top of my head began to tingle. And I started to hear a noise. It was guttural. It was very deep. It was a *natural D*," she recalled, with the ear of a musician with perfect pitch. She listened to the harmonics for a few moments.

"As the sound continued, I don't know how to explain this other than to go ahead and say it: I popped up out the top of my head," Pam said, looking for my reaction. "It felt like a suction cup at the top of my head, popping. And then I was looking down at the body, and I knew it was my body, and the odd thing was, I didn't care. It was wonderful."

From her vantage point just above and behind Dr. Spetzler's shoulder, Pam said she could see the entire surgical team. She wondered why they needed twenty people in surgical gowns to operate on her. At first she thought she was hallucinating but realized she felt too clearheaded to be on a drug-induced trip.

"My hearing was better than it is now, my vision better than it ever was, colors were brighter, the sounds were more intense. It was as if every sense that I had ever known—and add on a few—was perfect."

Pam's attention was drawn to the source of the natural D: an instrument in Dr. Spetzler's hand that looked like a dentist's drill.

"It was an odd-looking thing," she said. "It looked like the handle on my electric toothbrush. And there was a case—it all kind of freaked me out because it looked like my father's toolbox, like his socket-wrench case. And there were these little bits in there so it looked like he was doing home improvement and not brain surgery."

This was a Midas Rex bone saw—and Pam's was a near-perfect description of the saw and its blade container. At that point she noticed the other doctors midway down the table.

"It looked like they were doing surgery on the groin area. I heard a female voice say, 'Her arteries are too small.' And Dr. Spetzler—I think it was him, it was a male voice—said, 'Use the other side.' I'm thinking, *Wait a minute, this is brain surgery!* I did not know what they were doing. I was quite distressed, but about this time I began to notice the light."

At that moment, Pam's out-of-body experience ended and her journey "into the light" began. Michael Sabom, the doctor who analyzed the surgery from the medical documents, believes this may have marked the moment when her heart stopped and the brain-stem monitor flatlined. Pam's near-death experience contained common hallmarks: she saw a pinpoint of light that grew bigger and bigger, felt a pull toward the light, and then she heard a familiar voice.

"It was Grandmama," she recalled. "And I went to her. And with her was my musical uncle David Saxton," Pam's mentor, who had died of a massive heart attack years earlier. They looked young, she said. They shimmered as if they were wearing coats of light, and soon she spotted "a sea of people and they were all wearing the light."

"I remember asking, 'Is God the light?' And the communication was, 'No, He's not the light. The light is what happens when God breathes.'

"And I thought, I am standing in the breath of God."

My mind caught on the word "breath." The connection between breath and spirit dates back at least two thousand years. The Hebrews called it *ruach*, the Greeks called it *pneuma*, and what they meant was the spirit of God. When Jesus appeared to his friends after the Crucifixion, He *breathed* on them and said, "Receive the Holy Spirit."[5]

It occurs to me that perhaps this metaphor, like DNA, has been passed down through the generations, not because it is poetic, but because it is true: whether that breath arrives in death or in life, in practiced meditation or unbidden surprise, in first-century Jerusalem or twentieth-century Atlanta, that is what standing in the presence of God feels and sounds like—a wind that penetrates the heart, a breath that transforms a person and her world at a cellular level, a spirit that robs her of words but leaves peace in their stead.

Pam stood for a few moments in the breath of God. She yearned to go deeper into that light, but was stopped and told she needed to return. Her uncle, David, escorted a reluctant Pam back to the operating room.

"There I was again, with David, looking down at the body. Only at this point, that thing looked like a train wreck. It looked like what it was—dead. I did not want to get in it, I didn't even want to look at it, and now my uncle is reasoning with me. He says, 'Sweetheart, it's like diving into the swimming pool. Just dive in.'"

She protested, and then her uncle began reminding her of all her favorite things—her favorite food, her favorite smell, her favorite birdsongs—Pam's connection to the world.

"And I'm looking down and the body jumped. There were people around the gurney and the body jumped"—as they restarted her heart with a defibrillator. "And I thought, Okay, you know what, they're electrocuting that thing, I'm not getting in it.' Then my uncle *pushed* me! And I hit the body, and I heard the title track to the Eagles album *Hotel California*. When I hit the body the line was, 'You can check out anytime you like, but you can never leave.' And the body jumped again. That time I was *in* it and I *felt* it. And I opened my eyes and I saw Dr. Karl Greene, and I said, 'You know, that is really insensitive!'"

Pam laughed. "He told me I needed to get some more sleep."

At first she thought she had been hallucinating. But the next day, Pam met the cardiac surgeon who had commented on Pam's small femoral vein near the groin. From her angle, Pam had not seen the doctor's face during the out-of-body experience.

"I recognized her voice and I mentioned it. She looked at me kind of funny."

When Pam returned to Phoenix for her one-year checkup, she told Dr. Spetzler what she had "seen," including "doing the electric paddle thing" at the end of the surgery. And he said, 'Oh no. That didn't happen.' And he looked a little relieved, and frankly I was as well."

"Why relieved?" I asked.

"Well, if that part is wrong, maybe the rest of it is as well. Maybe it was just a hallucination," she said

"So I came back and told my doctor here. And he said, 'No, I'll check my records but I believe that they defibrillated twice'—which would make sense because I saw it once and felt it once. And sure enough, he confirmed it. And he called and talked to Dr. Spetzler, and Dr. Spetzler said, 'You know what, I wasn't in the operating room at that point.' "

Eventually Dr. Sabom conducted an exhaustive investigation of Pam's story. He obtained her records, including a timeline of the surgery and transcripts. He confirmed the conversation about her small veins, the description of the Midas Rex bone saw and its case. He confirmed the defibrillation, the number of doctors and nurses, even their position around the operating table.

I asked Pam's neurosurgeon, Robert Spetzler, how he explains Pam's perceptions.

"From a scientific perspective, I have absolutely no explanation about how it could have happened," he replied.

What about the drugs, or neurotransmitters, creating hallucinations?

"Those are suspect, but not in this setting," Spetzler said. "You can have patients who become hypoxic"—who have too little oxygen in the blood; in that situation, patients may experience hallucinations—"where you see yourself transported up into maybe a corner of the ceiling and you're looking down on things. But in virtually every one of those settings, you have a warm body which is missing something—

either not enough oxygen, or it's metabolically missing something—or it is feverish. So that you can imagine all sorts of neurons firing in an unorganized fashion and it would give you an explanation.

"Here you have the opposite," Spetzler continued. "Here you have the neurons in a depth of a sleepy state, of a suspended animation, that makes it very hard to think that it's from active neural transmission."

"So in your opinion," I said, "what does Pam's case say about consciousness, and whether it can be separate from the brain?"

"It comes down to the metaphysical," the neurosurgeon reflected. "It comes down to the soul. It comes down to whether you're religious and believe in these things. I think it is the ultimate arrogance for anyone, whether they're a scientist, or anyone else, to say that something can't be. I accept Pamela's account, although I have no explanation of how it could have happened."

Hallucinations and the Last Gasp of a Dying Brain

Gerald Woerlee believes he can explain it.

"It's a total load of rubbish," he said with a laugh.

The Australian anesthesiologist and author of *Mortal Minds*[6] is one of the feistier debunkers of near-death experiences. When I reached him by phone, Woerlee told me the mind simply cannot perform when the brain is disabled. Pam Reynolds's case, he said, crumbles under scrutiny.

First, he said, Pam awakened from the anesthesia when the surgeon began drilling her skull. Hence the "natural D." In other words, she was *conscious* right up to the moment that her heart fully stopped, and during this semiconscious period, she underwent her out-of-body experience. She "floated" out of her body, he explained, when the bone saw vibrated her muscle spindles. These are the movement sensors associated with your muscles. You may have learned about these sensors when you were a child if you ever pressed your arms against a doorframe, stepped away, and felt your arms floating upward. Woerlee theorized that when the bone saw vibrated, the muscle spindles contracted and suddenly Pam felt herself moving upward until she was perched above the operating table.

Alternatively, Woerlee offered, oxygen deprivation could be the culprit: When Pam's brain was gasping for oxygen, it began suffering from "hypoxic convulsions," or epileptic activity, an electrical storm in the brain. Woerlee contended this stimulated the part of the brain that located Pam in space, and suddenly, her brain told her she was not on the operating table but up near the ceiling, looking down.

Many neurologists embraced the epileptic-seizure explanation, and indeed there appears to be a link between temporal lobe epilepsy and out-of-body experiences.[7] Neurologists in Switzerland have even located a spot in the brain that sparks out-of-body experiences—and it is, of course, in the temporal lobe.[8] British skeptic Susan Blackmore argues that the brain creates models of reality based on sensory impressions, and when the normal perception is jarred—when the heart stops beating, for example—the brain draws on what it can: on memories and imagination.[9]

In this case, Woerlee said, Pam's addled brain created a "veridical perception" of her resuscitation. Perhaps her memory drew from what she had noticed as she was rolled into the operating room, or it could have drawn from previous memories, such as watching television shows like *ER*.

"But her eyes were taped shut," I protested, "and what she saw was *accurate*."

"Easily explained," Woerlee chuckled. Take the Midas Rex bone saw. "What she heard was a sound very similar to a dental drill, and a dental drill is something that everyone born in the 1950s understands and knows all too well. Most of them have four rows of lead in their teeth to prove it."

He argued that Pam's mind naturally created an image of the Midas Rex that looked like a dental drill. And she could imagine what an operating theater looks like from television and movies and perhaps from personal experience.

Okay, I said, not convinced but ready to hear more. "How could she hear conversations with those ear speakers in?"

"Have you seen a lot of these kids wandering around with earplugs?" he asked. I had noticed by now he favored the Socratic method.

"You mean iPods?"

"Yep. And the volume turned up to maximum? And nattering to each other?"

"Uh-huh," I said.

"Precisely," he said, as if that nailed the coffin shut.

"But they say these ear molds emit sound at ninety decibels, which is like a jet plane taking off," I observed.

"Yeah, it's all very well, but if you've got an iPod playing at maximum volume and you can hear your mate talking to you, then you're not going to tell me they can't hear each other. And that's basically the situation here. They make a lot of talk about these [ear molds] being totally sound occlusive, et cetera. But it's rubbish. She *heard* this cardiac surgeon."

The evidence suggests Woerlee is demonstrably wrong. The electrodes in Pam's brain stem had stopped showing any response to the ninety-decibel clicks, meaning she was not hearing anything, period. But Woerlee's insistence is revealing: no amount of evidence would budge him from his assumptions.

Next I asked about the oxygen-deprivation argument. Wouldn't her brain be too starved for oxygen for her to see or hear or form memories?

"No," Woerlee replied. There had to be enough blood in her brain to sustain consciousness "because this is proven by this event."

"Hold on," I said. "You're saying, if she had been unconscious, she couldn't form memories. Therefore she had to be conscious—because she formed memories. Isn't that circular reasoning?"

"Consciousness is a product of brain function. Period."

I couldn't resist baiting him one more time. "Some people believe that reductionist science is too narrow, that it has to come up with a new paradigm."

"*Everything* I say can be proven," Woerlee retorted. "*Nothing* they say can be proven. And that's the difference."

I relate this dialogue not because of its " 'Tis not," " 'Tis so" quality, but because this is the paradigm that near-death researchers must shatter in order to make their case. Woerlee and other mainstream scientists may be correct, and physiology as we understand it today may explain all. But when they hear Pam's accurate descriptions of the operating room, I wonder, aren't they just a little bit curious?

Sitting in her touring bus, I asked Pam if she felt like William James's white crow, living proof that consciousness is independent of the brain.

"I'm convinced it is," she said. "But it happened to me, so I sort of have no choice but to believe. I'm not a brain surgeon, so I can't speak to you on a scientific level of how it works. But in my mind, there's no way that consciousness is kept and recorded only in the brain."

She reflected a moment. "Every once and a while I wonder if it's just some kind of big farce."

Or, perhaps, evidence of another reality.

Before I left Pam's bus, one question nagged at me. I wondered if she had returned from the edge of death with a physiological marker in her brain, similar to the monks in Richard Davidson's research, or the nuns and Pentecostals in Andrew Newberg's studies. I asked Pam if her neurologist had found anything unusual about her brain.

She looked surprised at the question.

"Yes, my EEGs—my brain waves—are extremely slow," she replied.

"Even when I'm asked to do difficult mental functions, my brain waves act as if I'm at rest. But I can do the function and do it accurately."

"How do your doctors explain that?" I asked.

"They don't."

I got up to leave. I asked if I could take her picture outside. She laughed.

"I don't do stairs well, baby."

As I looked back, I snapped an image of Pam Reynolds's life: trapped in a broken body, yet savoring her husband, her children, her grandchildren. Walking with a cane but freer than most people I know. A 100-watt woman living in a dimly lit bus.

And the Blind Shall See

Pam Reynolds's story compelled me to seriously reconsider the assumption that the mind cannot function without the material brain. Vicky Bright's story took me one step further. Could it be that we possess *spiritual senses* that lie dormant until they are called into action?

At one-thirty in the morning on February 2, 1972, Vicky Bright (now Noratuk) had finished her shift as a pianist and singer at a small restaurant in Seattle. She was twenty-three years old and owned no car. With considerable misgiving, she accepted a ride from a couple who had enjoyed a little too much to drink. The wife was chattering away when the husband, who was driving, mentioned he was seeing double. Suddenly, they began to weave across the road. At the bottom of Queen Anne Hill, Vicky heard a sickening squealing of wheels. Their van spun and crashed into a wall, and then, in slow motion, Vicky sensed herself being flung out of the vehicle and dragged along the road.

"I saw the pavement," Vicky told me in an interview thirty-four years later. "I was up above, looking at my body on the ground." Her world went blank, until she found herself hovering near the ceiling in an operating room at Harborview Medical Center in Seattle. Below Vicky lay a woman on a cart—tall and slim, her thick waist-length hair shaved in places like a Mohawk. Vicky thought she knew the woman, but was not quite sure.

"I saw the people working on the body," she said. "There was a male and a female, and they were doctors. The male doctor was saying, 'It's a pity that she could be deaf, because there's blood on her left eardrum.' And the female doctor said, 'Well, she may not even survive. And if she does, she could be in a permanent vegetative state.' That really upset me.

"And I floated down closer to the female, and I tried to communicate with her. Then I saw my wedding ring, and I thought, *Oh my God, that must be me they're talking about.* And I thought, *Am I dead or what?*"

Vicky can perhaps be forgiven her confusion. This experience marked the first time she had ever seen herself. Or seen, for that matter. Vicky Bright has been blind from birth. Like many other premature babies born in the 1950s, she was placed in an incubator when she was born at twenty-two weeks, and the incubator destroyed her optic nerve.

"Some people say, 'Don't you see black?' No, I don't see anything, because my eyes are atrophied and the optic nerve is dead. So I've never seen light. I've never seen shadows. I've never seen anything, ever. Ever, ever."

That night in the operating room would be the exception. After she failed to draw notice from the female doctor—trying to grab the woman's arm and watching her hand pass through it—Vicky felt herself move "up through the ceiling and I was out up above the buildings and the street."

"Could you see the buildings and the street?" I asked, curious whether she could distinguish forms.

"Yes."

"Can you tell me what you saw? Like, did you see a car?"

"I saw several cars. They were going down the street. It was in the middle of the night. I saw buildings, I saw the street. But I had trouble discerning what things were—because, of course, I've never seen. And it was really terrifying."

She giggled. "But then, it was kind of neat, because I didn't have to worry about bumping into anything."

One could argue that all of Vicky's "observations" could be explained by assuming she was partly conscious during her out-of-body experience, and her hearing painted the images in her mind. She heard a male and female voice. The female doctor confirmed the next day that the doctors had discussed blood on her eardrum—an *auditory*, not visual, detail. Vicky's ears could have detected them shaving her hair and assumed it had been cut short in places. And her description of her wedding ring—white gold, with tiny orange blossoms around the diamonds—proved nothing since she had no doubt touched it thousands of times.

But if any part of Vicky's account is true, if she did "see" for the first time, it is the kind of evidence that smashes paradigms. On that operating table, Vicky found herself with a radically different way to perceive, a new "spiritual" sense that leapt into action the moment her brain was disabled.

That new sense is what Kenneth Ring calls "mindsight."

"The mind sees," he explained. "It's not the eyes that see. It's like a spiritual sight or a spiritual awareness."

Ken Ring, who is professor emeritus in psychology at the University of Connecticut, began investigating near-death experiences in 1977. Occasionally he stumbled across stories about blind people suddenly seeing objects for the first time.

"I felt if this was actually on the level, this would be a dramatic demonstration of the brain-mind split," Ring said, referring to the idea that the mind can operate independently of the physical brain. If he could find a case in which a blind person could accurately describe the environment, as verified by other people, "that really would be—I won't say a clincher—but a very strong argument for the authenticity of these experiences."

Ring never found his airtight case, but he did come to believe in a sort of spiritual perception. He tracked down thirty-one cases of blind people who had reported near-death experiences and inquired whether they had visually accurate (veridical) memories of those experiences.[10] Eighty percent reported visual perception—and of those respondents, two-thirds were blind from birth. Of his thirty-one subjects, fourteen reported an out-of-body experience, in which they claimed to visualize details in the operating room, their bedrooms, or other physical settings.

None of these stories makes a perfect case that consciousness continues when the brain has shut down. Still, the fact that all these subjects described the same perplexing type of vision does raise a startling possibility: somehow, Vicky Bright and others seem to have been catapulted into a new level of consciousness, where they found new resources to understand reality—in their case, sight that perceives in finer detail than ordinary vision. And the special case (of the blind) might hint at a more general principle: perhaps an encounter with death catapults you and me and anyone with all five senses into a different sort of perception of the universe. Being sighted, we cannot imagine what that is, any more than Vicky Bright could imagine seeing her shorn hair and her wedding ring before she glimpsed them in her out-of-body experience.

On any given day, you may be perfectly content to putter along in "normal waking consciousness," as William James had it, affixed to this familiar reality by your name, family, job, the opinion of others, your worldview, your bad knee, and your preference for strawberry ice cream.

"These are all things that keep you in this zone of ordinary waking consciousness," Ken Ring explained. "But if you nearly die, if you fall off a bridge, if something happens that shocks you, then [ordinary consciousness] falls away and you are for a moment aware of something else, something greater, something truer."

You perceive, he asserts, with "a kind of spiritual sense."

I was jolted by the language, not because it was foreign but because it wrapped around me like a soft, familiar sweater. Raised a Christian Scientist, I had learned to assume we have spiritual senses that perceive the spiritual realm—God's universe—just as our eyes and ears capture the physical world. Or, as Mary Baker Eddy, the founder of the religion, put it: "Sight, hearing, all the spiritual senses of man, are eternal. They cannot be lost."[11]

I had been taught that there is a close connection between material and spiritual existence, that out bodies and brains are the human counterpart to a spiritual identity—and that a *transcendent* realm can be perceived by spiritual senses. As I listened to these stories, I wondered if spiritual senses could be triggered not just by prayer and meditation but by drugs or seizure, by lucky happenstance, or by a close brush with death, as Vicky Bright and Pam Reynolds claimed.

It is one thing to swallow doctrine as a child, another to find that the very ideas you discarded as an adult might in fact be legitimate. It unnerved me. But I found that these ideas were not so strange after all. Consider, again, Aldous Huxley's "reducing valve." The author (and connoisseur of LSD) proposed that drugs, spiritual exercises, hypnosis—and, he might have plausibly added, death—can open up the valve to let us perceive at least a sliver of "Mind at Large." He defined Mind at Large as "something more than, and above all something different from, the carefully selected utilitarian material which our narrowed, individual minds regard as a complete, or at least sufficient, picture of reality."[12]

True, the strange idea that an invisible reality penetrates the material world, this bizarre notion that our minds might be able to operate independently of our brains—these are mere hypotheses. At the same time, these hypotheses about mind and brain are more than fantasy or speculation. They spring from life experience. These stories are everywhere, accessible. They are low-hanging fruit, and you need only step into your backyard and pluck them. And while people's experiences are not as conclusive as a DNA match, they are *something*, and they point toward a reality beyond the horizon of this world.

Dr. Bruce Greyson, a psychiatrist at the University of Virginia who has studied near-death experiences for thirty years, has noticed a rare but extraordinary phenomenon that can occur when people are slipping from life to death. As their brains shut down, they often enjoy brief and clinically inexplicable recoveries in the final moments of life. People suffering from dementia become lucid. Those with Alzheimer's disease recognize family members after years of confusion. Schizophrenics become clearheaded.

Even though this happens very rarely, he told me, "the fact that this ever happens at all is inexplicable if we equate minds with brains, because dying people's brains do not correct their structural or chemical derangement" when they are dying.

"When the brain is not functioning well," he reflected, "the mind becomes free to function."

And, perhaps, free to explore another, spiritual, realm.

Now let's travel all the way down the rabbit hole. Say, for the sake of argument, that these people who could see and hear when their brains were not functioning were not lying. Say their consciousness was not trapped in the brain. Does this unleash the outrageous idea that consciousness—your life and mine—survives death? Scientists summarily dismiss this idea as so much rubbish. I would not even raise it—except that now neuroscience is getting into the act.

CHAPTER 10

Are We Dead Yet?

The fragile yellow paper is undated, but my mother and I figured it referred to the winter of 1939. Mom was enrolled in a finishing school just outside of New York City. She was eighteen. My grandmother was forty-one. One bitterly cold evening, Mom received an urgent phone call. Granny had fallen deathly ill.

When Mom arrived at Granny's bedside, she found her mother slipping in and out of consciousness. Granny was attended by two Christian Science practitioners and a medically trained Christian Science nurse. They kept a fierce prayer vigil for nearly three days, but on Saturday morning, Granny "slipped out of this world and moved through a brief space of darkness," as she would later write, in a written testimony that my mother found after Granny's death. According to the people by her bedside, "the eyes did not close but a film closed over them and all activities ceased."

But not for Granny. Granny was on the move. "I had passed the portal called death," she wrote. "There was no fear and no anxiety. I seemed to be walking or going some place. I was conscious of the fact that I had left the world and those dear to me just as much as if you walked out of a room and closed the door behind you. After walking for a time, the light seemed to be breaking through and everything seemed to be getting much lighter when suddenly a light that I have never seen anything like before broke before me, and I was completely surrounded by a brilliancy that blinded me, so that I could hardly see. . . . A voice spoke to me and said, 'Go back, you are needed there.' As a soldier obeys a command spoken to him without a question, so I obeyed this command."

An hour passed, and the friends next to Granny's bed continued to hold vigil, unaware of Granny's subterranean travels. Suddenly, "to the astonishment of them both, I opened my eyes wide" and began to speak.

"I heard my own voice talking, and this is what was said as they took it down. . . . 'It is wonderful.' 'It is beautiful.' 'The darkness is all gone, there just isn't any more darkness at all.' 'There is no death. You don't have to die.' Turning to the [friends], I said, 'You never have to be afraid again.' "

At that moment, Granny threw off the covers. "I'm hot!" she declared, and rose from the bed, brushed her teeth, and asked for some breakfast.

When my grandmother returned from this ethereal voyage, she gave no external sign of the internal rewiring—at least, not in her personality. She remained an efficient, independent woman, one who would soon defy the convention of mid-century America and file for divorce. She continued her "healing" as a Christian Science practitioner. But the near-death experience instantly shifted her view of life, as if she switched from a magnifying glass to a telescope. Her prayer life, it seemed, increased in intensity by orders of magnitude, and she became somewhat famous for her healings.

"After that, she brought about the most remarkable physical healings that I know of personally," my mother recalled.

"Do you know why?" I asked.

"Yes. She said to me, *'Never be afraid.'* She said the whole experience taught her that there is nothing to fear, and that everything is love. That love was the light. There was no death, and so you need never be afraid."

Back then, no one discussed such far-fetched phenomena. Even Mom was unaware of the details until she inherited her mother's papers, four decades later.

Beginning in the 1970s, particularly with the advent of books like Dr. Raymond Moody's *Life After Life*,[1] thousands of people rushed forward to recount their journeys to the brink of death and back. These accounts contained many of the same elements: the white light, the beauty, the peace, the sense of viewing one's body from the outside. Skeptics believe that these similar phenomena reveal not a universal mystery pointing to eternal life but a common neurological response to the brain shutting down. But tell that to my grandmother, tell that to anyone else who has touched death—and tell that to increasing numbers of scientists.

If out-of-body experiences represent a pivotal battle in the war over consciousness—that is, whether your mind, will, and identity are only the expression of brain chemistry—then near-death experiences represent the battle to come. The questions posed by these experiences are too spooky for most mainstream scientists. Does your soul survive death? Is this life a shallow introduction to eternity, and death a sort of summer recess between kindergarten and first grade? Are we like stick figures on a chalkboard who live in two dimensions—until one day we pop off the board and find a three-dimensional world with not just length and width, but depth as well? And does that next dimension leave physical fingerprints on those who have traveled there and back?

The Perfect Death

A brief brush with death is the Hail Mary of altered consciousness: it's risky, but if it works out, you score big. Visions, peace, and serenity, light and love, unity with all things, dramatic personal transformation: everything that psychedelics or temporal lobe epilepsy, meditation or spontaneous mystical experiences offer, you can find in one near-death experience. Which is not to say you should stick a fork into an electrical socket; most people do not return from the edge of death. But it does explain why I found the three hundred people gathered in Houston such an embarrassment of riches.

The 2006 conference of the International Association for Near-Death Studies marked a watershed moment for the movement. Hosting the conference was M. D. Anderson Cancer Center in Houston, one of the premier cancer hospitals in the world. After years toiling on the outskirts of science, this endorsement by a world-class medical center tickled the near-death folks to no end. The participants at the conference—the subjects and scientists alike—were like a tour group watching the changing of the guard at Buckingham Palace, and suddenly being invited inside to dine with the queen. Yes, I thought as I settled into my comfortable seat in the hospital's ultramodern auditorium, near-death experience has *arrived.*

Near-death experiences have always been with us, but not until recently have they been mentioned in polite company. It took the book *Life After Life,* which was first published in 1975, to crack the dam of embarrassed silence and allow all those repressed stories to gush out. And gush they did—with such force that scientists began to divert them into different categories: the stages and elements of near-death experiences,[2] who gets them,[3] why they get them, how people are transformed, and various psychological, physiological, and neurochemical explanations, or, rather, the failure of all those explanations to explain away the caress of death.

In the past three decades, scientists have conducted more than forty studies of nearly 3,400 near-death experiencers. Articles have appeared in such peer-reviewed journals as *The Lancet*, *Nature*, *Brain*, and *American Journal of Psychiatry*, as well as the subject's flagship publication, *Journal of Near-Death Studies*. Most of these studies are fairly rudimentary. The problem comes down to money: Who is going to fund this research? As a result, a small band of researchers has toiled with barely enough bread and water to sustain them, compiling stories and statistics, until they have now amassed a fairly large body of research.

If one were looking for the *Mona Lisa* of near-death experiences—perfect and mysterious, each stroke masterfully executed—I would point to the story of Edward Salisbury. I met Edward on the first day of the conference in Houston, a tall, lanky man partial to checked shirts and black leather vests. His salt-and-pepper beard was neatly trimmed, his black hair sat flattened docilely against his head. When he spoke to me, he gazed kindly but unswervingly into my eyes. Those eyes held me hostage, as did his story.

On December 29, 1969, Edward was driving to see his fiancée in a new Firebird convertible. He was twenty-six, a young executive at Coca-Cola, driving too fast for the wet, winding roads of Atlanta's suburbs. His car hit a curb and suddenly—*BANG*—he felt as though he'd been hit by a swinging door from behind. His car plowed into a tree and in the next instant, Edward was "merging" with the tree, going up its arteries and popping out on top, twenty-five feet above the road.

"As I looked around—it was like I was looking off a balcony—I said, 'That car looks like *mine*,'" he recalled. "I looked a little closer and there was this body slumped over the steering wheel. And I realized, Hey, that's my body! If that is my body, then who am I? How did I get here? And in the next instant I was sucked away into a cascading, swirling wormhole of transformation."

Edward found himself in a beautiful bright presence, "like walking out of a movie theater in the daytime," and sank into the warmth as into a steamy bathtub. He soaked in the peace and joy for some time. Then he began to investigate the scenery.

"It's as if I had walked into the Library of Congress. There are these giant tomes, and I looked around, and every question, every concern, every thought I had, was answered as quickly as I formed the thought. Questions like, 'Why was I here?' 'How come my father was so upset with such-and-such event in his life?' And all my questions were answered."

Eventually Edward exhausted his questions. He spotted a majestic figure sitting on a throne—someone, he said, who bore a striking resemblance to Charlton Heston in his movie role as Moses. This was God the Father, apparently, and God patted His left thigh and said, "Come on," like Santa Claus in the department store.

"And in the next instant, I am in His lap, with His arm supporting my back, and He is pouring down this smiling love, it was in His eyes—the love I see in mothers when their baby is given to them at birth," Edward recalled. "And God said, 'This is where you always are, if you would only know it.' "

Edward recalls basking in God's gaze, until God pointed down to His feet. There Edward saw images, like so many pictures from an album that had spilled on the floor—memories spanning from his childhood to just a few days before the accident. He relived the memories from a 360-degree perspective, not only what *he*, Edward, had done and thought and felt, but how his actions affected others. He saw the girl in elementary school who had a lisp, and recalled teasing her because he wanted to be accepted by the other children.

"So I mimicked her. And I not only recalled the experience, I felt how *she* felt. I experienced the consequences of my behaviors."

Edward paused, looking truly stricken. "As I recall it now, I want to cry with how cruel I was and how painful and hurtful that was. And I turned again into God's eyes, and said, 'I'm so sorry.' And the response was an unconditional, loving acceptance that said, 'It is neither good nor bad in the greater sense. Are you through?' "

In this way, Edward marched through his entire life, recalling his bad acts and his good. The time in high school when he rescued a girl from drowning at the beach, the moments a few days before his accident when he had fudged on an expense account. Finally Edward had picked up the final snapshot: his life review was complete. Sensing that they were done, and that he might be asked to leave the peace and return to his wrecked body, Edward announced that he did not want to return.

"I said, 'Let me just put this pile of pictures back in order, and just send a message to Mom to let her know I'm okay, not to worry,' " he said. "I had a momentary thought—an earthly attachment—and Boom! I woke up in the hospital bed with my mother rubbing my feet. It was two weeks later."

Edward's story is a masterpiece. Not only does it include every element of a near-death experience, but his description precisely echoes the stories of other people at the Houston conference: the out-of-body experience, viewing his crumpled body with puzzlement and painlessness. The perfect, infinite knowledge he received in the "Library of Congress," as if tapping into a conscious intelligence, all-knowing and all-seeing. The total acceptance and lack of judgment by God or the light, depending on one's religious doctrine. The life review, in which the experiencer lives through the experience from at least two sides: his actions and their impact on others. The personal transformation that the traumatic experience sparks. Edward, for example, gave up his business career to become "a minister, a yoga teacher, and my wife's chauffeur." Finally, the return to his body, which was triggered by a connection to his life, to unfinished business and relationships.

Because I had been exploring the brain in the throes of "mystical" experience, I saw patterns that I might otherwise have missed. These events were not unique to dying. People suffering from temporal lobe epilepsy recalled having visions, hearing words or music, feeling the presence of the sacred. Those who ingested psychedelics—especially those who were in the final stages of cancer—recounted full-blown dramas that included death and resurrection, oneness with all things, an overwhelming love, the certainty that the death of the body did not mark the end. In both cases, when people returned from their drug trip or death trip, their fear of death had evaporated. Spiritual virtuosos in prayer and meditation reported not only visions (for some) but also an overwhelming peace and unity with all things and the certainty that there is more than this material world. Spontaneous mystics could have written the scripts for some of the near-death experiencers; but mainly they felt an overwhelming love that had nothing to do with theology or religious denomination. It seemed that, with the possible exception of temporal lobe epilepsy, encountering the "other," or the "other side," radically transformed the experiencer, as if rewiring his priorities and even his personalitiy. Every person I met in Houston testified to that fact.

For several days, I heard stories like Edward's, all passionate, all spoken as divine revelation, all compelling. What I did not hear was airtight evidence that these mystical voyages did in fact take place—that they were something other than tricks played by a dying brain. The near-death-experience advocates desperately need that proof. For their claim challenges one of the foundational principles of modern science: that your consciousness depends entirely on your brain, and when the brain shuts down, so does your identity. These survivors who refused to die fervently believe that consciousness exists apart from the brain, and, by extension, that there exists another dimension beyond what we can see and hear and touch in this material world. That claim will not triumph without a fight.

The Brain's Operatic Demise

Gerald Woerlee is spoiling for a fight. These "delusions," the Australian anesthesiologist said, are "easily explained" from both a narrative and neurological point of view. It's simply what a brain does when it is dying.

Woerlee argues that when a person's heart stops beating, he will appear to be dead because he lacks enough oxygen in his brain to move or to see, even though the brain stem, which controls the most basic functions like breathing or swallowing, remains active. Because this brain is not functioning normally, Woerlee says, it sets off a cascade of neurological events that explain every part of the near-death experience. When the prefrontal cortex malfunctions, you have a sense of calm, serenity, peace, joy, and painlessness. When the primary motor cortex malfunctions, you can't move. When the postcentral gyrus malfunctions, you can't perceive touch or sensation. When the parietal cortex malfunctions, you can't perceive where your body ends and the universe begins, making you feel at one with the universe. When the angular gyrus malfunctions, together with muscle spindles, you can believe you are moving or flying.

I can accept that someone could look lifeless and still remain conscious at some level. Who has not heard about people who emerge from a coma recalling the sounds and events around them? But the common experiences—the tunnel, the beings of light, the life review—how can these be explained away?

"The tunnel-and-light business is one of the easier bits to explain," Woerlee replied.

A brain in distress—that is, a brain gasping for oxygen—releases adrenaline, which widens the pupils of the eyes. And when the pupils dilate, he said, they can let in a hundred times the amount of light that pupils do under normal conditions.

"And so, if someone has had a shock or their pupils dilate due to illness, disease, fear, drugs, or oxygen starvation, that person will actually say, 'Hey, the room's getting lighter.' You look around and think, 'He's mad, the room isn't any lighter.' Of course it isn't! His pupils are wider, that's all."

That may explain the infusion of light, I said, but what about the tunnel?

Simple, Woerlee said. It's a matter of blood supply. Most of the blood goes to the central part of the retina, the part that focuses on objects, and when oxygen supply dwindles—as it does during cardiac arrest—"the bits that fail first are those that are furthest away from the center," he said. "And so, gradually, you get a narrowing of your peripheral fields, until all you've got is a central spot of light, and if your pupils are wide open at the time, you see light pouring into a tunnel."

It seemed to me that Woerlee made a persuasive case in describing generic experiences such as the light and the tunnel. But the cinematic experiences unique to each person—how would he explain that? How could generic brain activity manufacture an image that means something only to the dying person: seeing Uncle David in robust health, for example, or feeling pangs of guilt for ridiculing a shy little girl in fifth grade?

The brain gasping for oxygen, Woerlee said, is a little like the OK Corral—a desperate, frantic shootout in the brain. In fact, it resembles an epileptic seizure.

"And that stimulates all sorts of parts of your brain to do all sorts of wonderful things," Woerlee argued. The brain in distress stimulates sections of the brain deep within the temporal lobe,[4] like the hippocampus and amygdala, which house your memories and emotions. "And in general, when you stimulate the amygdala and the hippocampus, you get memories of people, of events, of sounds, music, sometimes even deities, and also memories of past events—flashback, life review, and then your visions of relatives."

"So all those visceral, meaningful events," I asked, "they're just in your memory bank, and they are roused when your brain is having a seizure?"

"Precisely."

True, Woerlee conceded, this analysis may sound like "a modern form of phrenology,"[5] the pseudoscience in which bumps and fissures in the skull were said to indicate personality traits. In fact, some neurologists dub it "neurophrenology." But he argued, "a veritable flood" of magnetic resonance imaging (MRI) studies during the last fifteen years supports his theories.[6]

This seems like a lot of heavy lifting for a brain in the throes of dying. It reminds me of those operatic death scenes in which the tubercular soprano trills away with perfect pitch and volume through her final aria.

Aware of these arguments, I sat down with Peter Fenwick, a neuropsychiatrist and fellow at the British Royal College of Psychiatrists. Fenwick specializes in treating epilepsy. He has never had a near-death experience himself but has researched it for the better part of twenty years.

"No epileptic seizure has the clarity and narrative style of an NDE," he said. "And this is because all epilepsy is confusional. Epileptologists all agree that one thing that near-death experiences are *not* is temporal lobe epilepsy."

What happens in a seizure, he told me, is that the electrical activity at that site *disrupts* the normal processing of the temporal lobe. When people report sensations from temporal lobe stimulation, they describe disjointed, isolated phenomena. Someone might say, "It feels as if I am leaving my body." Or, "It sounds as if I'm hearing a symphony." But these are isolated sensory phenomena, not coherent narratives.

Fenwick also dismissed the dying-brain argument. Twenty seconds after the heart stops beating, he said, a person is unconscious, period. You cannot argue that there are "bits" of the brain that are functioning.

"Everything that constructs our world for us is, in fact, 'down,'" he stated. "There is no possibility of the brain creating any images. Memory is not functioning during this time, so it should be impossible to have clearly structured and lucid experiences."[7]

Bruce Greyson, a psychiatry professor at the University of Virginia who is considered the father of modern research on near-death experiences, added that because of brain damage, a person should not be able to remember any experiences after his near death. "Even if you can establish that there is some residual brain function going on," he said, "that's not the same thing as saying there is enough integrated brain function to have elaborate thinking and memory formation during these procedures."

It's like saying that if an eight-year-old can pitch in Little League, then he can start for the Red Sox. An oxygen-deprived brain blurts out idiosyncratic hallucinations and leaves the survivor confused. But near-death experiencers tell coherent narratives and describe elaborate conversations with dead relatives, beings of light, or religious figures like Jesus. And even if those "memories" are not real accounts of actually meeting dead relatives in heaven but only a reconstruction from past events, that complex thought could not be formed while a person has only brain-stem activity or partial consciousness. Besides, the oxygen-deprivation argument cannot explain cases in which oxygen levels are normal upon death, as in a car accident.

Still, as I sat back and reflected on all these arguments, it seemed to me that the near-death-experience team has hit a couple of singles but not a home run. It is not that their arguments are unreasonable. It is that they are, at this point, speculation. How to directly test these experiences in, say, a brain scanner, is also problematic. Researchers can't really say, *Mrs. Brown, you don't have long now, may we just slide you into this MRI tube for the betterment of science?* Even if you could capture the experience in a neurological snapshot, what would that tell us? That various parts of the brain light up while Mrs. Brown is subjectively experiencing a near-death experience? We still would not know if the brain is *causing* the perception of something that is *not* happening, or leaving a *record* of something that *is*.

Bruce Greyson supplied what seemed to me the most honest and ultimately satisfying solution to this conundrum. I asked him if these near-death experiences point to a reality beyond our physical reality.

"There is absolutely no scientific evidence to make a compelling case one way or the other," he said. But after researching the edges of death for thirty years, he believes the evidence in favor of unseen reality is "impressive."

"We could be misinterpreting things, overemphasizing certain things. I would not be surprised at all if I'm wrong. But I don't think that's the case. I think the evidence strongly points in the direction of there being more than just this material world."

In thinking about two adjacent, perhaps overlapping worlds, I recalled something Pam Kircher told me at the Houston conference. Kircher is a physician at M. D. Anderson who specializes in helping patients at the end of their lives. Very ill patients train their senses on two different audiences, like an emcee who faces the audience on one side of the curtain, but occasionally pops his head through the curtain to see if the actors are ready. Kircher said when she started visiting dying patients, she noticed that they routinely talked with deceased relatives, the familiar "Aunt Sally," as it were. At first, she thought they were hallucinating, so she constructed a test.

"I would interrupt the conversation with Aunt Sally," she said. "I'd ask them their pain measurement on a scale of one to ten, or ask them what they had for breakfast. And they could tell me. They were polite; they would stop the conversation with Aunt Sally and tell me very logical answers to those things, and then go back and talk to Aunt Sally, who of course was much more interesting than I was."

A person who is hallucinating cannot be pulled back to reality, Kircher explained. "They're gone. They are not going to tell you what their pain level is. They're in another stage of reality. But my patients could be pulled back."

They have glimpsed behind the curtain to the stage she cannot see.

Dying in a Brain Scanner, Sort Of

The Holy Grail for near-death researchers is a physical marker, like a stamp in a passport that testifies that Mrs. Brown crossed into sacred territory and returned. In thirty years of focused research, scientists have never located such a marker. Perhaps a marker exists, perhaps it doesn't—but until recently, scientists lacked both the technology and the funding to even try.

Neurologist Peter Fenwick believes those markers do lie somewhere in the folds of the brain or the rhythm of its electrical current. Any major neurological event registers in the brain and then manifests itself in behavior. The brain images of people with post-traumatic stress disorder, for example, show cerebral changes.

"So it's likely that people who have a transcendent experience will also have changes in their brain as well," Fenwick speculated. "This is shown really because they then have changes in behavior. With post-traumatic stress, it's increased anxiety. In near-death experiences, it tends to be more social awareness, more spirituality, and so on. So these will in fact be accompanied by some cerebral markers. I'm sure we'll find them when we start looking for them."

Which brings us to the University of Montreal, where the hunt for a spiritual marker is in full cry.

Jorge Medina winced slightly as I shook his hand in the entryway of the University of Montreal Medical Center. We exchanged halting hellos—Jorge in his shy, stuttered English, his third language, after Spanish and French. I searched his face for some signature of trauma, and found wide brown eyes, a hearty black mustache, a face smooth and coppery and completely unmarred.

I unclasped Jorge's hand, and let my gaze fall to his forearm. There lay a tapestry of mottled brown-and-white skin, as shiny and inflexible as vinyl. His arm was a partial road map of his journey through the flames. Fire had left ninety percent of Jorge's body with third-degree burns, mercifully leaving his face unscathed.

"I'm sorry," I murmured. Now it was my turn to wince.

"No problem." He smiled, and we turned to the task at hand—one of the most controversial studies ever conducted at Montreal's illustrious medical center. We were about to scan Jorge's brain as he relived the moment he died.

Our guide was Mario Beauregard, a forty-something French Canadian neuroscientist. Beauregard was conducting cutting-edge research on the brain in mystical states.

Why, I asked him, would a promising young researcher risk his career by studying spiritual states and near-death experiences?

"Oh, that's easy," Mario had replied in his soft accent. He smiled shyly. "I'm a mystic."

During my visit to Montreal, Beauregard elaborated. "When I was eight years old," he said, "I had a kind of vision, and the vision became a certainty for me—that the brain was not the same as the mind and the soul. These things were different. And I decided then to become a scientist to demonstrate later on that this was the case indeed. That you cannot reduce a human being to a batch of chemicals and bones. And that became"—he searched for a word, *almost* getting it—"the motor, the starting point of the research that I'm doing right now."

Now, on a beautiful July morning in Montreal, this mystic was unlocking a thick door meant to stop the electromagnetic field generated by the fMRI scanner from leaking out of the room, and leading Jorge and me into a clamor that only very large machines can produce.

There we were greeted by Mario's lanky, handsome researcher, Jerome Courtemanche. Ignoring the din created by the MRI machine, he explained the process to Jorge slowly and carefully. They would take images of Jorge's brain in three states. Once he was lying comfortably in the brain scanner, Jorge was to relax and think of nothing in particular. This was the "resting state," which would be the baseline to compare against the other brain images. After sixty seconds, Jerome would tell Jorge through a microphone in the control room to imagine a light generated by a lamp. This was the "control state," which would theoretically activate the parts of Jorge's brain that involve memory and vision. After another minute, at Jerome's command, Jorge was to return to resting state. Then, a minute later, Jorge was to seek the "target state"—he must try to return to the "light," the luminous presence he experienced two years earlier, when he had died in the ambulance on the way to the hospital. In this way, the researchers could determine whether the mystical state of seeing the light differed from the mental state of seeing a simple lightbulb. They would repeat this six times. If the two states appeared identical, that would suggest there is nothing inherently special about mystical experience. It is a person's interpretation and not the event itself that infuses it with meaning—the way, perhaps, a sirloin steak tastes more succulent on the night your beau proposes marriage. The steak is not better than the one you had last Tuesday night but the context makes it so.

I watched from behind a plate of glass as Jerome led Jorge into a cold room. There, waiting ominously in the dim light, was the MRI machine. The brain scanner looked like a large oven, openmouthed and gaping for an object to be placed on the rack and inserted. I wondered if Jorge, already a victim of heat and fire, recognized the awful irony. If he did, he showed no sign: he merely stripped off his shoes, belt, watch, and any other metal object, and hoisted himself onto the gurney-like rack. Jerome fed him into the machine, a sacrificial offering to the gods of science, where enormous magnets would record blood changes in his brain.

Moments later, Jerome ordered the procedure to begin and a piercing ring filled the control room. It seemed unlikely than a yogi master could reach a transcendent state in this racket, much less a Mexican hotel worker who has never meditated a day in his life. I turned to Mario Beauregard, who was standing next to me watching through the glass.

"So, what do you think you'll find?" I asked.

"I think we'll see regions involved with positive emotions light up, of course, because it's a very positive experience for a subject," the neuroscientist speculated. "We should also see brain regions that are involved in self-awareness, because there's a change in self-awareness during such spiritual states. And also, the body representation of the subject shifts and the subject becomes less in touch with his body, so we might see major changes in the parietal regions of his brain. But," Beauregard said cheerfully, "that remains to be seen!"

Jorge Medina was the first of fifteen people to re-create his near-death experience in the brain scanner. Beauregard had no trouble recruiting subjects. A newspaper advertisement drew more than a hundred volunteers, allowing Beauregard to be selective. The subjects had to meet three criteria: when they had nearly died, they had lost consciousness and entered the "light"; they had returned with a transformed view of life; and they could still reconnect with the light.

In the control room, Jerome was saying, "Super, Jorge, we're done," and asking Jorge to rate how close he came to the light during his various target states, on a scale of one to five. Because Jerome has recorded the exact times Jorge was trying to achieve the target (near-death) state, he could later match Jorge's subjective scores with the images of his brain during those moments. Jerome beamed like a proud parent: on a scale of one to five, Jorge had reported mainly fours and fives, a surprisingly strong intensity for a man lying in a metal casket with ringing sounds as loud as ambulance sirens going off around him. Jorge emerged, grinning.

"How was it?" I asked. "Did you connect with the light?"

"Yes, yes, yes, I feel a little bit today like my accident," Jorge said. We sat down then, and Jorge described that moment as he lay in the ambulance, his body covered with third-degree burns.

"Peace—the peace that is like joy and like love," Jorge whispered. His eyes welled up as he described his brush with death as "a nice experience."

Jorge Medina had arrived in Canada seven years earlier, and began working a string of jobs as a cook and housekeeper in a hotel, sending money back to Mexico City, where his wife and children lived. On Sunday, August 1, 2004, he dragged home after a long shift at the hotel, fretful about his life, about money, about his son's impending marriage. He relaxed in the living room, beer in one hand, cigarette in another, until sleepiness overcame him. He trundled off to bed, only to be awakened later by black smoke billowing through the apartment. The living room had turned into a furnace.

"I decided I needed to put it out," he recalled, "and I began to get burned as I crossed the living room. I tried to grab the extinguisher, but I couldn't because it was so hot. I felt the flames on my body but I didn't realize I was on fire."

Jorge ran down three flights of stairs, a live ball of flames, and reached the street. Neighbors doused him. He lay on the street, waiting for an ambulance that would take twenty minutes to arrive.

"They raised me into the ambulance, and I began to feel separate from myself," he told me, reverting to Spanish to articulate the ineffable. "I felt like everything ended right there. And then I found myself walking toward a door, trying to open it and not able to do so. There was a very large window, with a lot of light in it. I think people were behind it. I felt like I needed to wait by the door for someone to open it for me. And someone said to me—well, it wasn't a voice, it was 'peace' that said to me—'Everything will be okay. You wait here.' I said, 'Okay.' I felt extremely calm the entire time. I didn't know anything of myself until I woke up three months later."

Could it have been morphine? I asked.

Jorge said he asked the doctors the same question, and they told him they did not administer morphine until he arrived at the hospital. Beauregard, who was listening to the interview, said that morphine produces a different, more fragmented experience.

When you met the light, I continued, was it a person, was it God, was it Jesus, what was it?

"At first I thought it was God or some image of God, but now I think we will never truly be able to grasp what God is," Jorge observed philosophically. "It can't be described. It was a light, and it was peace."

The doctors and nurses dubbed Jorge "the Mexican Miracle." He survived nine surgeries, eleven blood transfusions, and months of painful rehabilitation. He stopped drinking and smoking. His family and his Catholic Church moved to center stage in his life. He grew less anxious about his problems with family, work, money. That moment in the light rewrote his vision of the future, and not just in this world.

"I used to *believe* there was life after death," he said urgently. "I *believed* in God. But I lived like everybody—in between yes and no. Now I *know* there is something else. When I think about death, I think about how nice it is to be alive and to be with my family. At the same time, I don't worry about what's going to happen later. Everything will fix itself."

"Can you tell me what happened in the brain scanner today?" I asked.

"I simply tried to recall the experience, and I began to see the light. I cannot say I saw God, but there was a state of peace, and I left my body. Time and space became relative. Oh, one other thing," he said offhandedly. "Normally I have a lot of pain in my hand," he said, rubbing it gently. "And at that moment in the light, it didn't hurt at all."

Jump-starting Your Spiritual Life

Two years and fifteen subjects later, Mario Beauregard could talk about the brains of people who had touched death and returned. When he and I were observing Jorge in the brain scanner, the neuroscientist had predicted that Jorge's brain might look similar to those of some Carmelite nuns he had studied. Like Andrew Newberg at the University of Pennsylvania, Beauregard had conducted brain-imaging studies of people engaged in "centering prayer."[8] His subjects had all lived in the cloister for decades, where they spoke for but one hour a day, and their lives orbited around periods of this sort of meditative prayer: on average, they each had spent nearly 15,000 hours in prayer.

When Beauregard eventually coaxed the nuns from their cloister to the brain scanner, he was effectively able to shoot a movie of brains in mystical union with God. The images told a complicated story: areas associated with positive emotion became a happy cauldron of activity; the circuits associated with unconditional love also lit up brightly and the parietal lobes, which determine the subjects' physical boundaries—where they end and "God" begins—showed unusual changes in blood flow. Their brains seemed to be saying that the nuns felt themselves absorbed in something greater than themselves. In the end, Beauregard amassed sufficient visual evidence, in the form of brain images, to make the case that a mystical state was physiologically *distinct* from either an intensely emotional state or a resting state.

His research suggests that the brains of Carmelite nuns operate differently from the ordinary brain. In this, his conclusions matched those of Richard Davidson at the University of Wisconsin, with his Buddhist monks, and those of Andy Newberg at the University of Pennsylvania, with his Franciscan nuns, Buddhist monks, and Protestant Pentecostals. Spirituality, it would seem, does leave an indelible fingerprint.

Now Beauregard's question was: Would near contact with death trigger the same sort of neural activity? When he sat down and compared the brain reliving a connection to God in prayer with a brain reliving a connection to the "light" in a near-death experience, he made a remarkable discovery: a near-death experience unfolds in the brain in much the same way as a meditative union with God. It lights up the same areas and travels along the same neural pathways. One uses Google Maps, the other uses Yahoo!, but they visited many of the same points along the way.[9]

For example, Beauregard noticed that both for those who nearly died and for those who meditated, the part of the brain usually associated with "the subjective experience of contacting a spiritual reality" showed a spike in activity: the nuns sensed that they were in touch with God; the near-death experiencers felt they had contacted the "Light." Another part of the brain associated with overwhelmingly positive emotions lit up. The same occurred in an area of the brain that involves unconditional love, both romantic and maternal love.

"It's not surprising to see this type of activation during the NDE condition," Beauregard theorized, "because love is one of the key components of this experience."

Beauregard's research suggests that both meditation and death rewire the brain in similar ways. It comes as no surprise that Catholic nuns and Buddhist monks can essentially sculpt their brains, fine-tuning them to access another dimension of consciousness. After all, they engage in rigorous mental workouts every day, praying and meditating for hours at a time. But what about Jorge Medina, who had never meditated a day in his life, and before his accident rarely darkened the doorway of a Catholic church? His brain gave him access to a spiritual or altered state of consciousness, just as theirs did—but without the decades of training. It seemed to occur in an instant, jump-starting his spiritual life.

Or, as Beauregard put it: "It's like it will accelerate to a great extent the spiritual path for transformation."[10]

Beauregard discovered another startling similarity between spiritual virtuosos and the survivors of death. Not only did similar parts of their brains light up when they reenacted their experiences, but they could both manipulate their brain-wave activity, almost like a key, to open the door to a spiritual realm.

I watched this occur at seven p.m. on July 12, 2006. Mario Beauregard, his research assistant Jerome, and I squeezed into a cold, windowless twelve-by-eight-foot room, along with Gilles Bedard, one of Beauregard's subjects. The room was crammed with computers, some piled one atop another. Gilles sat on an ancient wooden chair, wearing a blue-and-red plastic cap with thirty-two small white buttonholes in it. Jerome bent his six-foot-three-inch frame over Gilles, filling the holes with gel and then meticulously attaching electrodes to his scalp. These electrodes would measure electrical activity in thirty-two parts of Gilles's brain—and, the scientists hoped, produce a brain-wave recording of a near-death experience, or at least a simulated one.

Gilles's near-death experience happened upon him on November 17, 1973. He was a nineteen-year-old with undiagnosed Crohn's disease. During his five-month stay in the hospital, he had wasted to sixty-four pounds. One night, after a priest had given him the last rites, Gilles found himself perched in the corner of the ceiling, looking down at himself, the doctor, his family, the room.

"And suddenly, I felt a call. . . . I just turned maybe ninety degrees, and there was a huge light in the back, and in front there were twelve beings of light," he recalled. "So I came upon them, and I said, 'What's happening?' They said, 'Well, you're not going to die. You must go back. You have things to do.' They were talking to me by telepathy."

"Were you afraid?" I asked.

"No. I felt very at peace. Then I asked them, 'What am I going to do?' And there was a silence, and then a beautiful sound came into me. It was outside of me, it was inside of me, I was part of a very peaceful, serene sound, but at the same time a very powerful sound. It was like"—Gilles blew out his breath slowly, pursing his lips to make a low whistle—*"the breath of the universe."*

It was an echo of Pam Reynolds's moment "standing in the breath of God."

Gilles recovered and, happily for Mario Beauregard, could revisit the "light" at will more than thirty years later. We settled Gilles in the acoustic chamber, then closed the door behind us with a quiet *thlunk*, the seal as tight as a submarine's. Beauregard and Jerome walked back to the computer, which was recording the undulating lines of thirty-two electrodes.

"He's already relaxed," Jerome said excitedly, as the lines on the computer rounded out from sharp spikes to small humps.

"The EEG is slowing down," Beauregard said, pointing to the screen. "This is good, very slow waves . . . theta waves . . . maybe it's a correlate of his experience."

Theta waves, Beauregard explained, are observed in people who are sleeping, or in a state of quiet focus, such as meditation. Beauregard said the average person cannot simply sit down and achieve this level of relaxation in two or three minutes. Over the next hour, Gilles sat in his soundproof tank, herding his thoughts toward various levels of consciousness. Beauregard and Jerome clucked fondly over Gilles's EEG recordings. Finally, at their command, Gilles plunged into the "light"—or at least, that is the story that the luxurious, voluptuous, sloping lines on the computer told. *Look at us! We're delta waves! We should be in a deep sleep, but really we're in another state of reality!* Well, that is what I imagined his waves were saying.

Beauregard later told me that some of his subjects had been able to slow their brain-wave activity to show delta waves—or states of deep sleep—and all fifteen had shown significant theta waves in their brains, which also occurs during meditation.

"What does that mean?" I asked.

"We know they've been transformed *emotionally* by the near-death experience," he said, "and perhaps they've been transformed *physiologically* as well. It's like there's a shift in their brain, and this shift allows these people to stay in touch with the spiritual world more easily, on a daily basis."

Beauregard explained that the University of Montreal maintains a database of the EEGs of thousands of people in the normal population. Most people at resting state exhibit the beta waves of normal waking activity, darting here and there, always in motion, thoughts swinging like monkeys from tree to tree.

"What we've seen with our subjects is that they present much slower waves in their brains than normal people"—especially theta waves of meditation and delta waves of deep sleep.

"Yet they are not impaired neurologically, so that their brain is not functioning slowly," he said excitedly. "And it's possible that it's a biological marker of their near-death experiences—that they are changed by the experience, even physiologically."

I recalled that Pam Reynolds, whose brain was shut down for more than an hour during a surgery, had told me that her brain-wave activity was "extremely slow," even when she was performing complex tasks. Richard Davidson at the University of Wisconsin found that his monks possessed a different brain rhythm—very fast gamma waves—which he said were a unique marker of meditation. And then there was Andrew Newberg at the University of Pennsylvania, puzzling over the fact that the Tibetan monks in meditation, the Carmelite nuns in prayer, the Pentecostals speaking in tongues, all showed the same neurological quirk: one thalamus was more active than the other.

Perhaps a close encounter with death has the same effect as a psychedelic drug. Perhaps it swings open the door to a spiritual realm, instantly accomplishing a state that usually requires untold hours of prayer. And once you have come in contact with God, or the Light, or the ground of being, or another reality, the experience leaves a fingerprint on your brain—that metaphysical stamp in your passport that proves you've traveled to another reality.

Beauregard said it is possible to test this hypothesis by looking at the EEGs, the brain scans, the immune systems, and the DNA of people before cardiac arrests and then after them. He wants to work with surgeons who perform the kind of surgery Pam Reynolds received: the "standstill procedure," which takes one's brain off-line for forty-five minutes or longer. That would certainly improve the odds of capturing a near-death experience. Beauregard mused that researchers might be able to determine whether the brains of those who experienced the "light" really did operate differently from the brains of ordinary, mildly spiritual people like me. And if they did, the question became, did the NDE brain start functioning differently *before* they died—suggesting that some people are born with the brain wiring to allow them mystical experiences? Or was it *after* they died—suggesting that their brains were altered by the experience? Beauregard has already proposed such a study, and hopes with the fervor of a true believer that he will one day know the answer.

"Of course," he added with a sly smile, "the spiritual world, the *source*, has to be willing to play that kind of game for us to be able to do this."

Yet even this sort of study cannot prove that the soul survives death, or that life continues after our hearts stop beating. It would only demonstrate that the brain is recording an event, either a real encounter or a hallucination. And there is another problem. Even if everyone from

Richard Dawkins to the Dalai Lama agrees that *something* occurs in the boundary between life and death, no study can prove the content of near-death experiences, because no objective scientist can follow the dying person into the light, shake hands with Aunt Sally or Jesus, and return with a clipboard full of notes.

I outlined my concerns to Mario Beauregard. He nodded. Some things, it seems, must be taken on faith.

"So what do *you* think, Mario?" I asked. "Are we're hardwired to connect to God?"

"Yes, of course, we are neurologically equipped to be able to connect with God," he said. "It's not the same thing as saying we are hardwired to do that, because if that would be the case, we would have seven billion mystics living on the planet. But is the equipment there to allow that to happen? Oh, yes."

Of course, Mario Beauregard is a mystic.

Every person I interviewed who had traveled to the brink of death returned with a new definition of God. I had first noticed this when I talked with people who had enjoyed spontaneous mystical experiences, and I saw the pattern repeat with those who experienced other transcendent moments as well. I realized that after encountering the "Other," people no longer clung to religious distinctions. If they had identified themselves as Christian or Jewish before, they might still attend church on Sunday or synagogue on Saturday, but they no longer believed their faith tradition could make a claim of exclusive truth. They were like witnesses to the same God, but from different angles. Or, think of God as the head of a multinational corporation. He controls several subsidiary companies, each with its own president: Jesus heading up Christianity, Moses overseeing Judaism, Muhammad guiding Islam, the Buddha launching his own belief system, and on and on. But take the elevator up one level, above the religions that try to make sense of the spiritual world, and you find the "Other" or "Light" or "Source"—that is, the CEO who presides over the whole enterprise. Now, I am not saying I agree with the view that all of the world's great religious traditions hold, at their root, the same view of the nature of reality. I am simply reporting what spiritual adepts told me.[11]

The "God" of mystical experience fits much more comfortably with the one described by Albert Einstein and other great scientists than the neat deity proclaimed by the average cleric. As I inched my way forward, watching and listening to the stories, I found myself wrestling with my own definition of deity, and in the light of morning I would walk away with a new name for God.

CHAPTER II

A New Name for God

THE SCIENTISTS WHO HAVE ARRIVED at God's doorstep have not, on the whole, taken a route illuminated by simple religious faith. They have traveled there by evidence, by looking at very small things, or very large: exploring DNA and quantum particles, or the puzzles of the universe. And in the blueprint they see the hand of God.

I use the word "God" to mean the organizer of the atom and the universe, but other terms fit the description. Albert Einstein, who dismissed a personal God but was persuaded by the evidence to assume there is a cosmic one, concluded there is a transcendent source of rationality in the world. He called this source "a superior mind," or "illimitable superior spirit," or a "mysterious force that moves the constellations."[1] When speaking of the order of the universe, Einstein stated, "That deeply emotional conviction of the presence of a superior reasoning power, which is revealed in the incomprehensible universe, forms my idea of God." Einstein once explained that he did not believe in a supernatural being who answers prayer, but that did not exclude the existence of the Almighty. "Every one who is seriously involved in the pursuit of science becomes convinced that a spirit is manifest in the laws of the Universe—a spirit vastly superior to that of man, and one in the face of which we with our modest powers must feel humble."[2]

Einstein's vision of "God" presaged the words of later scientific luminaries. Consider theoretical physicist Stephen Hawking, who spoke of "the mind of God." It is not enough to think of the order of rules and equations that make life possible, he said. One must contemplate "what it is that breathes fire into the equations and makes a universe for them to describe."[3] Asked why he believed the universe exists, he responded, "If you like, you can define God to be the answer to that question."[4]

The pioneers of quantum physics, who peered into the atom and were astonished at the mysterious world they beheld, saw God in much the same way. "God is a mathematician of very high order," observed quantum physicist Paul Dirac, "and He used advanced mathematics in constructing the universe."[5] They saw no conflict between science and religion. Physicist Max Planck, for example, wrote that religion and science are "fighting a joint battle in an incessant, never relaxing crusade against skepticism and against dogmatism, against unbelief and superstition," concluding with an odd sort of rallying cry: "On to God!"[6]

Even Anthony Flew, a lecturer at Oxford and one of the twentieth century's most renowned atheist philosophers, was converted in 2004 by the logic of God. Nearing his eightieth birthday, Flew abandoned the atheism on which he had built a career because, he asserted, intelligent life could not arise randomly. It must have been sculpted by the hand of a Creator.

"The journey to my discovery of the Divine has thus far been a pilgrimage of reason," Flew wrote. "I have followed the argument where it has led me. And it has led me to accept the existence of a self-existent, immutable, immaterial, omnipotent, and omniscient Being."[7]

Recently I listened to Francis Collins deliver a lecture to a group of skeptical scientists about his own journey from atheism to Christianity.

Collins is one of the country's leading geneticists and the longtime head of the Human Genome Project at NIH, which created the genetic map of a human being.

Collins argued that belief in God, "while not provable, is more plausible than atheism." He arrived at this conclusion by following scientific arguments, not ignoring them. He identified several "pointers to God" that nudged him toward the notion of a Mind that created the universe and, more remarkably, life. One pointer is the fact that there is something—a universe with people—rather than nothing at all. Another pointer is the Big Bang: Collins argued that if there was a beginning to the universe, this required "a Creator not bound by laws of space and time" who lit the fuse.

A third argument for God is "the unreasonable effectiveness of math," he said, suggesting "there might be a Mind with a mathematical bent beyond anything we can imagine." Collins cited the "Anthropic Principle," which posits that the universe is "finely tuned" to produce and sustain life—or, as physicist Freeman Dyson put it, "the universe in some sense knew we were coming."[8]

Scientists noticed that a handful of "constants" in the universe, like strong and weak nuclear forces, must have the exact numerical values they do for the planets to exist, not to mention intelligent life to develop to the point where you can read this sentence.

"If you tinker with any of the constants," Collins told his audience, "the universe could not support the complexity needed for intelligent life." If gravity were weaker, for example, matter would fly apart and planets could not form. If gravity were stronger, the Big Bang would have been followed by the Big Crunch, and life could not have developed.

Collins came to believe in "a God who loves math, wants a universe, wants complexity so that beings could evolve into intelligence and wonder if there is more." From there he went on to explain why Christianity makes sense to him. (For our purposes, that is a different story, more in the realm of personal preference than scientific induction.)

"God" may not be, as the atheists have it, a *delusion*—but perhaps a *conclusion* driven by the math of the universe. The infinite intelligence that maintains the planets in their orbits and tailors the molecular composition of air to each breath we take—this intelligence is not the figment of a narrow fundamentalist mind but the property of the most rigorous scientific minds. This is a God who makes sense to me, a defensible God, and one who has a starring role in a new batch of scientific experiments.

God 1.0

Imagine stripping God of all His imagery. Gone is the throne, the beard, the Michelangelo painting of a majestic Being nearly touching Adam's finger with the spark of life. Gone, too, are the stories of a God who intervened, who favored a certain people, who assumed the physique of a man. This stripped-down version would be the sum of his attributes, which would include infinite information, an omnipresence that fills all space and connects all atoms, a taste for mathematics that keeps the planets in their orbit, and the power to do so. This is a God who might appeal to the concrete thinking of a scientist. I came to think of this as "God 1.0"—God minus the love and the narrative history.

Larry Dossey calls this God "non-local mind." Dossey, a doctor and author, coined the term in his 1989 book *Recovering the Soul.*[9] It bears more than a smart scientific ring. "Non-locality" is a staple of quantum mechanics, and one of the spookier aspects of physics. For Dossey and others on the edge of science, "non-local mind" refers to a consciousness that defies the bounds of space or time.

"Perhaps the place to start is to say that non-locality is simply a fancified word for infinitude," Dossey told me one luminous day in July 2006. "If consciousness is non-local, then it is infinite in space and time. If something is infinite in *space*, it's omnipresent. If it's infinite in *time*, it's eternal, or immortal. So you can see that from the get-go there's theological dynamite that's hooked up with this idea of non-local mind."

Larry Dossey and his wife, Barbie, had invited me to their rustic home on a mountainside overlooking Santa Fe, New Mexico. Dossey, a tall, lean Texan with a mane of white hair parted in the middle, began his career as an Army battalion surgeon in Vietnam, then moved to internal medicine for fourteen years before turning to the pen. When I met him, he had already authored ten books on healing and spirituality.

When I, a stranger, wrote at the last minute requesting an interview while I was passing through New Mexico, Dossey set aside the entire afternoon before he and his wife left for vacation. It turned out to be one of his last mobile days that summer. Two weeks later, as the couple rode out of the Wind River Mountain wilderness area in Wyoming, Dossey was thrown from his horse. He broke his back, fracturing the twelfth thoracic vertebra. Because a helicopter could not land anywhere nearby, the Vietnam vet walked through the night nearly ten hours to civilization, in excruciating pain.

Dossey has taken as much incoming fire from his scientist colleagues as he did from the Viet Cong. What he proposes is a revolution in science. If non-local mind were merely the equivalent of the Divine Watchmaker, who created the universe and then let it tick away on its own, the idea would unsettle fewer of his colleagues. But Dossey's claim is far more galling: he suggests that this non-local mind interacts—has a relationship, even—with a person's individual, *local* mind, in the same way that many Christians or Jews believe that God interacts with people. According to Dossey and a growing number of scientists (along with most of the American population), this cosmic consciousness permeates our world and soaks into our human affairs.

Think of your "local" mind as your personal computer. You can keep files and write documents that no one else can access. "Non-local" mind is like the Internet: it contains enormous amounts of information, shared by billions of people (potentially by everyone on earth), and is always available for you to access with your individual mind.[10]

Dossey theorizes that your consciousness shares qualities with non-local mind, that the local and the infinite are "two sides of the same coin." This may seem far-fetched until you begin reading the mystics or practicing meditation or listening to anyone who has enjoyed a profoundly spiritual experience. They witness to being "at one" with the universe and God, feeling the boundlessness of the infinite, and experiencing "the divine within." And if there is a dialogue, so to speak, between your mind and the larger non-local mind, then it follows that your mind could do things that modern science says is impossible, such as impact other minds or know things that you simply should not know.

"One of the things that scientists have had a difficult time doing is to imagine how consciousness might behave non-locally," Dossey observed. "That it might exert its effects beyond the individual brain and the body, as in the stuff that parapsychology deals with, like ESP, clairvoyance, this sort of thing. And, we must add, intercessory prayer, which has always been a claim of all major religions. So it has been verboten to suggest that this actually happens, that the mind can behave non-locally, because every respectable scientist is dragooned into the notion that, by definition, *that can't happen*."

Dossey leaned forward in his chair and gazed at me with his penetrating eyes. "Here's the problem: The data hasn't gotten the message. The data doesn't know it's not supposed to occur."

He recalled one data point that grabbed his attention and prompted him to revisit his concept of time and reality. He was in his first year of private practice and had recently moved from "agnosticism bordering on atheism" to a yen for Buddhist writings.

"I had a dream one night in which I saw a patient undergoing certain tests," he recalled. "I even saw how the outcome of the tests would manifest. I went to my office the next day, and the scenario played out in camera-like detail. This was non-locality of consciousness manifesting outside of time—*precognitively*, as we say, anticipating something in architectural and camera-like detail before it even happened.

"Now, my experience doesn't amount to very much," he conceded, sitting back in his chair. "But you couple this with the record of the human race as far back as we know, and even surveys in contemporary America would show that precognitive dreams—dreaming something before it happens—are one of the commonest so-called parapsychological experiences that human beings have. One option is to say that the vast majority of Americans are nuts. There could be a case for that. But I don't think that we're all fooling ourselves when we talk about these non-local manifestations of consciousness."

I want to step back from seeing the future in "camera-like detail," to the more commonplace "gut feelings" that most of us have felt. A mother waking up certain that her son is curled up with stomach flu on his bunk bed at summer camp. A man sensing that his wife has just been in a car accident. Almost everyone carries around an anecdote like that. To test the hypothesis, simply tell one of these stories at a dinner party and wait for the dam to burst.

Dean Radin's Entangled Minds

Such stories present circumstantial evidence of a stripped-down God 1.0, an infinite intelligence that knows past, present, and future, is everywhere, and can communicate with us through our own (local) minds. But science always demands a mechanism, in this case, an explanation of *how* these uncanny, some would say spiritual, phenomena occur, including prayer and healing.

Dean Radin has a hypothesis: We have "entangled minds." Radin, who authored a book by that title,[11] is a senior scientist at the Institute of Noetic Sciences (IONS), a research organization founded by former astronaut Edgar Mitchell to rigorously study the "inner space" of consciousness. Radin could be a middle-class computer whiz straight out of central casting—bespectacled and balding, with a pencil-thin mustache, his voice a tad nasal and his speech hyperarticulate. Radin loves his numbers—he performs all the statistical analysis for the research projects at IONS—but he possesses an original mind that is two parts science and one part imagination.

As a child, Radin practiced violin several hours a day and eventually played professionally. His brain, as I understood from my exposure to virtuoso minds, was physically molded by the hours of practice. He was a loner who devoured hundreds of books in the public library, beginning with fairy tales, moving on to science fiction, and finally to mythology.

"Early on I was attracted to the notion that there were multiple layers to any story," he recalled, gazing out his window at the mountains surrounding the IONS compound, a wilderness paradise some forty miles north of San Francisco. "I never forgot that virtually anything that people are presenting to you, even in science, has multiple levels of meaning."

This intuition—that there may be a hidden reality—led Radin to *entanglement*. The idea of entanglement is this: when you delve down to the subatomic or quantum level, particles remain connected even when they are apparently separated. Albert Einstein called these connections in quantum theory "spooky action at a distance."

When Einstein was alive, entanglement was only an idea that was predicted by mathematics, but it had not yet been demonstrated in the laboratory. That would begin to happen in the 1970s, when researchers first started to explore whether the predicted properties of entanglement could be observed in the laboratory. In a groundbreaking study in the 1980s, French scientist Alain Aspect and his colleagues experimentally caused two photons, or light particles, to become entangled. When a property of light—such as spin, position, or momentum—was measured in one of the particles, the "twin" particle instantly showed the opposite property. What was especially spooky was that distance between particles did not matter. Even though the twins were more than thirty miles apart, they behaved as if they were still connected. They were entangled.[12]

Radin is quick to point out that entanglement has been shown only at the subatomic level, and that we human beings are much bigger than that.[13] But since people and things are composed of subatomic particles, Radin argues, entanglement may suggest that *everything* is interconnected, even people. We are not billiard balls on a pool table that occasionally bump into each other. We are part of a fabric woven so tightly that pulling one thread alters the whole tapestry. Or, try this: Reality is like Jell-O: Poke one side of the bowl and the green stuff on the other side jiggles. According to Dean Radin's entangled reality, if that "poke" is an event—say, a car accident—information about that event could pop into someone's head miles away.

"The mystery with all psychic phenomena is a mystery of how information gets from one place to another," he explained. "So if you hear the phone ring and somehow instantly you know who's calling without answering the phone yet, how did the information get into you? Where did it come from? Of all the infinite number of places where the information could have gone, why did *you* pick it up?"

"Which reminds me of a story my sister-in law, Katherine, told me just the other day," I volunteered. "Katherine's great-grandmother was traveling from Los Angeles to Chicago by train, and suddenly in the middle of her journey, she jumped up and said, 'My husband has just severed his forefinger. I must get off the train and go home.' When she arrived in Los Angeles a couple of days later, lo and behold, her husband was missing his finger. Can you walk me through what happened?"

"It's not as though the pain suddenly found her in a train," Radin hypothesized. "They were connected. They always had an open channel. The moment something happened to him, she was aware of it instantaneously because of that continuous connection. Now the interesting question is: Why didn't everybody else jump up with the same information? And the answer is roughly like this: In an interconnected space, everything truly is interconnected with everything else. If you think about what it would be like, it would be overwhelming. If you were able to sense exactly what is happening to everything else throughout the universe, it would be such a cacophony of information all the time that you couldn't think straight. So I believe that we have evolved in such a way as to exclude almost all available information, except the stuff close to you, in front of you here and now."

This dovetailed with the metaphor of Aldous Huxley, who proposed that the brain is a "reducing valve" that strains out all extraneous sensory information. He suggested that the valve occasionally opens—during a psychedelic drug experience, for example—to admit normally inaccessible information. It is like a voice an octave above our usual range of hearing, a light just outside the visible spectrum. Some people with temporal lobe epilepsy claim to hear and see another dimension, as did mystics from the beginning of recorded history. Here Dean Radin was suggesting that other things—important events, people, relationships—have the same ability to break through the veil of ordinary perception.

"If you are at cocktail party, and someone says your name across the room, you tend to hear it," Radin said. "Why do you hear it? You're not consciously listening for it. But you have maybe ten things you're constantly scanning, unconsciously, and these are a little bit easier to get into your conscious awareness than all of the other things you don't care about. So, by the same token, it's almost as though we are in a universe-sized cocktail party all the time, and we have the ability to have our attention placed subconsciously on the things that are of interest to us. So in that case, your sister-in-law's great-grandmother had one of her attentional filters subconsciously attuned to her loved one."

Dean Radin and other scientists around the world are running tests to determine whether this kind of connection is coincidence or a measurable phenomenon, whether we are disconnected billiard balls or dynamic beings living in an entangled tapestry that some would call God's universe. This amounts to declaring civil war on normal science, since most scientists insist that one person's thoughts or consciousness cannot extend beyond the brain, and certainly not far enough to affect another person's vital signs.

Yet a small avalanche of experiments suggests that we may in fact be connected in this way. Researchers at universities around the world—from Stanford to the University of Edinburgh—have conducted more than fifty studies to test the hypothesis, and found a "small but significant effect" of one person's thinking upon another person's body.[14] Some studies using EEGs showed that when the "sender's" brain-wave activity changed, the "receiver's" brain-wave activity followed within milliseconds.[15] Other researchers employed brain-scanning technology and found that when the sender tried to relay an image to the receiver, parts of the receiver's brain that handle visual images were activated.[16]

In one intriguing experiment, Jeanne Achterberg, a psychology professor at Saybrook Institute in San Francisco, and her colleagues gathered eleven shaman healers on the Big Island of Hawaii.[17] At random two-minute intervals, each healer prayed for a close friend as the friend lay in a brain scanner. When Achterberg analyzed the brain scans, she was astonished at the results. During the intervals when the healers were sending their prayers and intentions, the partners' brains "lit up" in the areas that are usually activated when someone does something or says something in response to a stimulus—possibly, she theorized, the healer's thoughts. The odds that chance alone accounted for the partners' brains lighting up in this way were 7,874 to one.

Mainstream scientists rightly question these studies. They ask whether the statistical analysis was done properly, whether the findings have been replicated or are just a fluke, whether the researchers had a stake in a particular outcome (a flaw not limited to these off-Broadway scientists, by the way, but is endemic in scientific studies). These are legitimate questions, even though the studies I reviewed were rigorously conducted and published in peer-reviewed journals.

I include them here because materialist scientists have not *explained* things that people experience routinely—such as gut feelings, premonitions, and certainly not the power of prayer—except to dismiss them as coincidence. Had they presented compelling theories for these everyday phenomena, I would not bother exploring alternative explanations. Since they haven't, we are going to take a brief walk on the wild side—through the parapsychology ghetto, a neighborhood that most scientists drive around. But like Harlem, parapsychology is enjoying something of a renaissance.

The Bonds of Love

Dean Radin and his colleagues' most recent experiment mirrors the one that Alain Aspect conducted with particles of light. In that experiment, the French researcher connected two particles and then separated them, only to find that they continued to behave as if they were still connected. Radin and his colleague Marilyn Schlitz, who directs research at the Institute of Noetic Sciences, tested the phenomenon on a larger scale: instead of bonded photons, they studied bonded people.

Radin and Schlitz selected thirty-six couples who were willing to see if their emotional bonds translated into physical ones.[18] During the experiment, each partner in a couple was placed in a separate room in which they could not talk to, see, or communicate with the other in any way; in fact, one partner was sealed in a soundproof, electromagnetically shielded room. Each person was also wired to allow a computer to record various physiological measurements: heart rate, respiration, brain-wave activity (EEGs), skin conductance (sweat glands), and peripheral blood flow. These are measures of a person's unconscious nervous system, which one does not control.

During the thirty-minute experiment, one partner would send ten-second bursts of focused loving "intentions" to the other at random times. The computer would measure each partner's nervous system. The intriguing question was this: When the "sender" transmitted his packets of love, would the "receiver" physically respond with a jump in her brain-wave activity, with a racing heart, or with sweaty palms?

"You have a long-term couple; they will both have a lot of physical proximity to each other," Radin explained. "They will be 'entangled' both emotionally and psychologically and maybe physically. And if they *are* physically entangled, you should be able to separate them, poke one, and see the other one flinch. And in essence that's what these experiments are looking at. They're looking at space separation under conditions where we don't know of any classical form of connection that would cause one to get poked and the other one to flinch."

That poses a revolutionary question right there, but then the "Love Study" added a twist. It made intuitive sense that couples who were "highly motivated" to connect would in fact perform better at affecting each other's vital signs. What sorts of couples, Schlitz and Radin asked themselves, are highly motivated to connect? One answer: sick ones. Of the three dozen couples in their study, twenty-two couples included one partner who had been diagnosed with cancer. Thus both partners should be "highly motivated" to see that their prayers or intentions had a physical impact on the ill partner.

I asked Radin if I might observe entanglement in action. He agreed, and on a gorgeous morning in March 2007, I caught a glimpse of Einstein's spooky action at a distance in the pulsating force of Teena and J. D. Miller's love.

Teena Miller breezed toward me, a study in pink: pink camisole and a pink silk blouse draping over her floral skirt. Her large straw hat shaded translucent skin and kept her red hair in place. It seemed a strangely feminine outfit for the Institute's rugged compound on the mountain, yet I could sense that Teena defined her own atmosphere. She did nothing halfway. A passionate liberal, she served as an executive board member of California's Democratic Committee. She joined rallies with little provocation and had accumulated an odd set of titles—indeed, she was a certified whiskey taster and, for some reason we did not explore, an honorary member of the International Order of Camel Jockeys. At fifty-seven, Teena had defeated cancer not once but twice.

A half-step behind trailed her husband, J. D. Miller, cupping her elbow in his hand, less to guide her, it seemed, than to simply touch her. At sixty-four, J.D. was trim in his khaki pants and royal-blue shirt, his easy smile framed by a white beard and mustache. J.D. was a CPA and financial planner, a Republican, a beta to Teena's alpha. Teena and J.D. had married nine years earlier, and to say they were "bonded" was an understatement. Teenagers, more like, taking every chance to embrace each other so intimately that I would avert my eyes. ("Excuse us," Teena said as she kissed J.D. on the cheek during one of these close encounters. "She calls it 'vitamins,'" J.D. explained.)

J.D. said he fell for Teena twenty years before he dared ask her out. In the interim, Teena married and gave birth to two girls, divorced, and raised the girls on her own. But J.D. never married. He was waiting, and when he bumped into Teena a decade ago, it took him all of two dates to make his intentions known.

Three years before I met the Millers, their relationship had come under scientific scrutiny in the "Love Study." The Millers qualified not only because of their palpable chemistry, but also because of a tumor in Teena's breast the size of her fist. She endured six months of chemotherapy, eleven operations, and six weeks of daily radiation before her doctor declared the cancer removed. Teena was in the middle of treatment when she saw a flier announcing a research project about "loving intentions." She picked up the phone and called the Institute of Noetic Sciences, eager to have another arrow in her quiver against cancer.

That is how Teena and J. D. Miller found themselves squirreled away in the mountains of northern California, wired to computers that would measure their vital signs and, like a Geiger counter, the seismic activity of their love. And it is how I found myself following the Millers, Schlitz, and Radin down a stone path through the woods to their laboratory nestled in the trees.

Radin led J.D. to a soundproof room, and sat him before a computer screen. Over the next half-hour, Teena's face would appear on the screen at random intervals, and remain there for ten seconds. During those times, J.D. would send his wife compassionate thoughts. The rest of the time he would think of anything but his wife.

As Radin gave J.D. some last-minute instructions, Schlitz was ushering Teena into the soundproof, electromagnetically shielded room. No noise, no signals of any sort, could enter or leave. Teena settled into the deep armchair and sighed contentedly, absorbing her environment: a plastic ficus tree, soft lighting, pale curtains, and a camera aimed at her face which would allow her husband to see her from the other room. Schlitz leaned down to affix electrodes to Teena's hand.

"So, this is measuring blood flow in your thumb, and this is measuring your skin conductance activity," Schlitz explained to Teena. Because they had dismantled much of the equipment after the Love Study was finished, they were measuring only two of the five bodily functions during this replication.

"Basically both of these are measures of your unconscious nervous system. And as you can see, your image is being projected by this little camera into the next room. So your husband will be able to watch you at random times during the session. You won't know when, so don't try to guess, because it's all random."

"Right," Teena said, leaning back against the headrest and closing her eyes as Schlitz closed the hermetically sealed door with a soft thud.

A minute later, the experiment began. After a few seconds, Teena's face appeared on J.D.'s monitor. We knew he was seeing Teena on the monitor and sending her loving intentions for those ten seconds; we gazed at the computer screen recording Teena's blood flow and skin conductance (that is, the perspiration on her hand). Those ragged or undulating lines would indicate whether Teena was perhaps unconsciously "responding" to J.D.'s thoughts projected from the other room.

"Notice how there's a change in blood flow?" Radin asked me excitedly, pointing at sine waves on the computer monitor. "A sudden change like that is sometimes associated with an orienting response. If you hear somebody whispering in your ear and there's nobody around, you have this sense of *What? What was that?* That's more or less what we're seeing in her physiology."

Thirty minutes passed, during which time J.D. sent thirty-six random packets of "loving intention" to his wife. The experiment ended. Radin slipped off to run through a quick computer analysis of the data. Schlitz walked to the sealed room and opened the door. I followed on her heels, intent on interviewing Teena and J.D. separately, before they could exchange their stories. I was curious to know whether their thoughts as well as their physiology matched in any way.

"What did you experience?" I asked Teena.

"Happiness. It went between my granddaughter and my husband. What goes through my mind is kissing my husband right near his ear, where the whiskers are. I love that spot. And that kept popping into my head, constantly."

"Were there any body sensations associated with it?" Schlitz asked.

"How do you explain bliss? How do you explain that in words, other than holding my granddaughter, or being hugged by my husband? It's just the best feeling."

I cornered J.D.

"What were you thinking about when you were sending your thoughts?" I asked.

"Some of the stuff I'm not going to share with you." He laughed, looking slyly at his wife across the room. "But others . . . well, she has a picture with her today that shows her with her granddaughter, and there's just a bond between them. It makes her feel so good, so I was thinking of those things."

Initially, I found it impressive that they had focused on the same two things—each other and Teena's granddaughter. But I quickly checked myself. *Of course* their thoughts overlapped: you would hardly expect them to be pondering presidential politics during a "love" study. In my head I heard the sonorous voice of Richard Sloan, the skeptic at Columbia Medical Center, scolding me: "Anecdotes, while interesting, are merely anecdotes. They are not evidence."

Equally intriguing was the story Teena told me while we waited for the results. I asked her if she thought that these connections that Radin and others had measured might simply be statistical flukes or chance happenings.

"It can't be," she insisted. "It's energy. I have always believed that, because I've been linked to certain people."

She paused. "For example, one day I knew something wasn't right with my daughter. She was nineteen, and I picked up the phone and I called a number. I don't know where the number came from. And this young voice on the other end of the phone picked it up, and I said, 'Hello.' And I immediately started talking. I said, 'I know you don't know me, but I believe my daughter is walking toward you at this very moment.' And he got very quiet.

"And he said, 'Describe her.'

"And I said, 'She's about five-two, she's very attractive, she's nineteen, she's got long dark hair . . .'

" 'What's her name?'

" 'Alison,' I said. Then I heard him calling, 'Alison!' "

Teena laughed. "She was in the parking lot at a gas station, her car had broken down, and that was the closest place she could get to. And when Alison heard him calling her name, she stopped, because she didn't know him, and he didn't know her.

"And the young man said, 'Your mother's on the phone. She wants to talk to you.'"

Teena smiled. "And that's what I mean when I say I'm linked with people."

Later, I phoned Alison to ask for her version of events. She recalled that in the fall of 1994 or spring of 1995, when she was a freshman at the University of Nevada in Reno, she was driving home from school to her mother's home near San Francisco. Her car began to break down near Sacramento. She managed to pull into a service station, and as she was talking to the mechanic, the phone rang. She heard him say, "Yeah, your daughter's right here," and he handed her the phone.

"Is your mother really intuitive?" I asked.

"Yes, but not like *that*. It was crazy."

Alison does not remember calling her mother from the service station or a pay phone beforehand—but she may have. She does not believe she gave her mother the telephone number at the gas station—but it is a possibility. I have eliminated one other possibility: the technology that would have allowed Teena to return the call automatically, without knowing the number—★69 or caller ID—was not available in California in 1995. As with so many other stories I encountered in my research, this one is not airtight: a skeptic will dismiss it and a believer will feel a shiver of recognition.

I turned to Marilyn Schlitz at IONS, who had been listening to the story and nodding.

"How can this be explained?" I asked. "Is this common?"

"Well, to have such a clear example, where you have a number which is seven digits, is pretty improbable, to say the least. Certainly the experience is widely reported, and we do see a lot of evidence for this kind of thing within the laboratory as well."

Of course, no matter how many strange stories you present, how this happens—whether it points to a quantum phenomenon or an old wives' tale—remains a matter of fiery debate.

Where Science and Spirituality Kiss

Moments later, Dean Radin marched back with a computer analysis of the results. He leaned over Teena, showing her a graph with lines that looked like a series of molehills. On a sheet of paper were the emotions that Teena had felt, translated into the language of physiology: squiggly lines that measured blood flow and skin resistance (perspiration). The lines showed that within two seconds of the time that J.D. began sending his "intentions" toward her, she became "aroused" (that is, increased perspiration) and then she would relax (or flush) when he switched off his intentions.

This one test does not offer definitive evidence that one person's thoughts affect another person's body. But the Love Study suggests there may be a concrete, measurable, physical connection between bonded couples.[19] When Radin and Schlitz analyzed the results of all thirty-six couples in their study, they found that when one person "sent" his compassionate intention to his partner, the partner's physiology mirrored his within two seconds.[20]

The correlations were not perfect, but they were powerful. And they did not happen by chance. The odds of getting the results they did by mere chance were 11,000 to 1 for all the bonded couples. And when you looked at the "highly motivated" couples—the ones who were facing cancer, who had something at stake—the odds against chance jumped to 135,000 to 1.

Now, let me be clear. I am not arguing that studies like this one explain reality. These findings could turn out to be a statistical fluke—although there have been some fifty other studies like this one that yielded similar results. I wouldn't be surprised if in five years, or twenty-five, scientists find wholly materialistic explanations for phenomena such as gut feelings or intuition. But I present the alternative studies here because they offer an explanation—something not pulled out of a magician's hat but based in quantum physics and tested under rigorous conditions.

"What we're experiencing here suggests that consciousness or mind or intention is *causal*—not only in our own bodies, but causal in the external world," Schlitz said.

When you pray for your child, or root for the Dodgers, or curse at your colleague under your breath, that has a measurable effect, she said. "We can't yet explain it, but this provides a revolutionary break in thinking about how we're connected, and how we're separated. I think we are smack dab in the middle of a paradigm shift."

As she spoke, a wild thought occurred to me. Perhaps quantum physics offers an explanation for prayer and healing. Maybe Einstein's "spooky action at a distance" is at work, in which prayers move from one person to another through Dean Radin's Jell-O–like reality, that closely woven fabric of reality in which all things are ultimately entangled. For some, that stuff is God, who does not so much intervene in our lives as provide the medium through which prayers move.

I was curious about how Teena explained the results.

"It's just the real strong power of energy between people who care for each other."

Finally, I thought. The missing element. My "God 1.0," the scientist's God, that intelligence that weaves the universe together, unifies everything in a web of consciousness, the God that resides non-locally, outside of time and space—that God had left me a bit cold. It was flesh and bones without life. What was absent was the *pneuma*, the breath of love.

Even though scientists on the frontiers of science rarely mention that line on God's résumé, everyone else who believes there is something more than this puts "love" first on the list.

It occurs to me that experiments like the Love Study are both very small and very large. We are talking about tiny variations in a person's sweat glands; it doesn't get much more incremental than that. And yet, perhaps Teena's fluctuating skin resistance swings the door open to a cosmic idea—that she is somehow connected across time and space to her husband and children, and if she is connected to them, then why not to every particle and person in the universe? That is what quantum entanglement—this Jell-O–like reality—is claiming. The implications are as vast as the heavens.

And while the terminology of "entanglement" and "non-local mind" may be twenty-first-century, the idea dates back to antiquity.

"This is the place where science and spirituality are converging," Schlitz observed. "You look at the great teachings of Buddhism, Hinduism, Daoism, a lot of the shamanic traditions, even in the Christian traditions, and you see this sense of unity and wholeness. And now what we're finding is that there is a justification for that sort of philosophical position from an empirical base, from a scientific base."

While this definition of God does little to fill my soul or feed my emotions—I gain little comfort from imagining myself in the arms of non-local mind, for example—it does go some way toward nourishing my intellect. As I was wrestling with my own concept of God, one that could survive the science I was absorbing, I suffered a small crisis that allowed me to put some of these ideas to a test.

The U.S. Postal Service and the Fabric of Reality

On January 30, 2007, a tiny error set off a chain reaction that threatened nearly a year of research. All year long, a young colleague of mine had been generously transcribing my interviews from various reporting trips, for a pittance. I could not have finished the book without her. Over Christmas, she had taken some minidisks (three-inch-by-three-inch square digital disks) to her family's home in New England and accidentally left them there. This particular set of minidisks included interviews from four trips (New Mexico, Canada, Florida, and Pennsylvania), which were essential for five different chapters—nearly half the book.

Her mother put the minidisks in a manila envelope and dropped them into a mailbox. On February 2, the envelope arrived in Washington with a large gash in it, minus the minidisks. When I learned this, the earth lurched to a halt, silence roared in my ears, and I was paralyzed by the implications like a mouse before a rattlesnake.

If you live on the East Coast and noticed a decline in your mail delivery service in the first half of February 2007, you can direct your ire at me. During that time, I enlisted the help of every U.S. Postal Service customer-relations officer between New Hampshire and Washington, D.C. By seven o'clock each morning I had methodically called the managers of mail distribution centers up and down the coast. I e-mailed them digital pictures of the minidisks. I called individual post offices, and enjoyed several conversations with the investigative arm of the U.S. Postal Service. By the end of the third day, the Postal Service officer who was in charge of my case remarked, "Ms. Hagerty, we have thirty people working to find your minidisks." I did not mention to her the dozen or so other federal employees I had called on my own.

I also took a step I had not considered in a decade. Christian Scientists, I have noticed, are strangely adept at finding lost things, possibly because they focus on God as "infinite Mind," the all-knowing and all-seeing. I called a Christian Science "practitioner" to pray for me.

"Hello, darling," she greeted me, with a warmth that almost reduced me to a puddle of tears. "How can I help?"

I explained the situation and then heard the familiar, steely resolve of a Christian Scientist in action. There followed a brief conversation, in which she made a series of declarative statements. Among them: "Nothing that belongs to you can be separated from you. Feel your oneness with the Source of intelligence." And: "You think *you* have covered a lot of ground in searching for this? God has already covered all the ground there is. Nothing passes His notice." And: "It's like the law of mathematics. It is always operating, and everyone is responding to this law. It's like gravity. You don't have to understand it but you obey it. Everyone is responding to this law of order, including the mailmen, the mail processors, the person on the street—everyone is under God's law of order."

You get the idea. At base, of course, this is not so far from the idea of non-local mind suggested by quantum physics—namely, that everything is connected in a fabric of intelligence, and we have access to this larger "mind" that is not limited by time or space.

The days stretched on and my disks made no appearance. Every moment that passed shaved down the chances of finding the minidisks before they were sent to a warehouse in Atlanta (the size of two football fields), or, more likely, thrown out as trash. But I stubbornly clung to this image of an intelligence that keeps an eye on everything; I had hit the "search" function and was now waiting for the results.

Nearly three weeks after the minidisks disappeared into a black hole at the U.S. Postal Service, I received a call from Dennis Panich, the manager of one of the distribution centers in Washington, D.C.

"I've got good news," he said. "One of my guys found your tapes."

I was struck dumb.

"Where were they?" I finally asked.

"They were in the bottom of a large orange container, along with a few other things." He laughed. "You know, I wrote a book myself, and I know how much work goes into it. So every day I showed my folks the photo of the minidisks and told them we were going to find those tapes. And we did."

When I visited the Institute of Noetic Sciences about a month later,

I related my story to Dean Radin. Radin does not believe in a traditional God, but he does believe we can access information by tapping into a larger, non-local mind. I asked him to explain the happy outcome from his perspective. I waited for the low, appreciative whistle of amazement that would make such a lovely scene in my book. No whistle came.

"Your story is mind-boggling in one way," Radin observed, "but I'm a little blasé about it now because I've seen it happen so many times."

I was both miffed and intrigued. How did the recovery work, according to his theory of entanglement? Remember, entanglement suggests that reality is like a fabric in which everything is woven together. Or, to continue with the Jell-O analogy: if you touch one side, the other side jiggles.

"Things appear to be separate," Radin said, "but with the right way of focusing your attention, they're not so separate. My suspicion is that you sort of imagined in your mind's eye that the universe is connected, and there was one piece of the universe—where the tapes were—that you *really needed.* And you created a network of attention, a lot of people's attention dedicated to finding this particular thing. There was an army of people out there who wanted to please you, they wanted to find those tapes, and so it's almost as if a momentary network was created that had a little bit higher valence to it, it shined a little bit brighter for you and all the other people, and this *group* attention became focused on finding the thing which is there."

Radin paused. "Everything is there," he said. "Nothing can be lost in a holistic medium."

I nodded, thinking silently that this "holistic medium" sounded an awful lot like "God."

My experience with the minidisks *proves* nothing, of course. To those who believe in miracles, God intervened to return my precious items. To those who think of the universe as a fabric of reality in which all is connected by a vast intelligence, I had tapped into information that was waiting for me all along. And to many others, I had caught a lucky break, nothing more. But let's be clear: Which way you view the universe and "God" is a matter of choice, not hard evidence. Hard science does not mean petrified science. And in this moment, science may be in the middle of a paradigm shift.

CHAPTER 12

Paradigm Shifts

THE MORNING OF JUNE 15, 2005, arrived way too early. For two weeks, my friends and I had been thinking hard all day and bantering long into the wee hours in pubs around Cambridge University—a grueling regimen for a forty-something crowd. But nothing could delay us from the morning's event, and so we marched from our hotel, past the famed Mathematical Bridge, to the Divinity School. Once inside the sleek building, we sat down at a long table of expensive blond wood and waited eagerly. We were anticipating the Fight of the Century.

It was a pointy-headed fight, to be sure, but we were a pointy-headed crowd: ten seasoned journalists[1] invited by the Templeton Foundation and Cambridge University to observe as celebrities in the world of science unspooled their ideas about biology, string theory, and multiverses. What tagged these presentations as unusual was the question underlying them: Could God retain a place in the intelligent man's world? Or, in this scientific age: Had God been reduced to a superstitious belief lacking any rational basis?

After eight days of lectures—and our polite but lethal grilling of the scientists—it seemed to me that God was losing. While many of the scientists claimed some sort of faith, generally Christian, they kept their spiritual beliefs in a sealed container where they would not contaminate the "real" work of understanding human consciousness and the universe we live in. I was witnessing a blitzkrieg of scientific materialism overrunning the quaint but untestable claims of God.

This irked me, especially when I realized that God could not win under the rules of twenty-first-century science. This was not Ali versus Frazier. This was the World Wrestling Federation. The decks were stacked, the outcome certain, the smack-down inevitable. The rules of this game—the paradigm of modern science—revolve around certain core beliefs. One of them dictates that scientists can study only what they can measure: the physical world and observable behavior. Try to investigate something that cannot be precisely measured—such as a spiritual experience that transforms a person's life—well, that's cause for immediate disqualification.

Another rule is the mind-brain paradigm: everything we are, see, feel, do, or think is a physical state, the electrical and chemical activity in three pounds of tissue called the brain. Mind, consciousness—forget about the *soul*—must be reduced to matter. It is a closed loop, excluding any notion of God or a spiritual realm.

But on that rainy morning in Cambridge I witnessed something extraordinary, akin to Dorothy spotting the little bald man pulling the levers of the Wizard of Oz. For only a moment, the curtain pulled back and we saw the fight for what it was: two *belief systems* duking it out.

John Barrow, a brilliant Cambridge mathematician, was speed-walking us through the hypothesis of a "fine-tuned" universe that is exquisitely and astonishingly calibrated to allow for life. He explained the concept of "multiverses," which posits that we live in one of 10,500 universes. Then he said, almost as an aside, "I'm quite happy with a traditional theistic view of the universe."

He might as well have dropped an anvil on Richard Dawkins's foot.

Dawkins is a renowned evolutionary biologist at Oxford University and possibly the world's most famous atheist, certainly one of the most militant. Two days earlier, Dawkins had delivered a talk that he believed would prove the impossibility of God, and which would later be published as a book called *The God Delusion*.[2] He had remained in Cambridge to hear the lectures of other researchers, particularly the world-class John Barrow. When Barrow, who turned out to be an Anglican, mentioned his belief in God, Dawkins began roiling with frustration like a teakettle about to blow.

"Why on earth do you believe in God?" Dawkins blurted.

All heads turned to Barrow. "If you want to look for divine action, physicists look at the rationality of the universe and the mathematical structure of the world."

"Yes, but *why do you want to look for divine action?*" Dawkins demanded.

"For the same reason that someone might *not* want to," Barrow responded with a little smile, as all of us erupted in laughter—except for Dawkins.

So there you have it. The paradigm is not a law, it is a choice: a choice to look for—or exclude—the action of a divine intelligence. The paradigm to exclude a divine intelligence, or "Other," or "God," to reduce all things to matter, has reigned triumphant for some four hundred years, since the dawn of the Age of Reason. Today, a small yet growing number of scientists are trying to chip away at the paradigm, suspecting that its feet are made of clay.

Some of them have patiently guided me through my own quest to understand the nature of God and my instinct that something does indeed exist beyond the reach of our physical senses. I have come to think of them fondly as "my" scientists: they are brave and passionate in their conviction that the materialist assumptions of modern science are not as sturdy as they look.

These scientists repeatedly invoked Thomas Kuhn's name. Kuhn, a historian of science at MIT, changed the world with his book *The Structure of Scientific Revolutions.*[3] Aside from popularizing the phrase "paradigm shift," he presented a new model for scientific progress. He argued that science proceeds not by steady accumulation of knowledge but (as one writer put it) by "a series of peaceful interludes punctuated by intellectually violent revolutions."[4] Those revolutions are "tradition-shattering complements to the tradition bound activity of normal science."[5]

Kuhn argued that scientists are not the skeptical, freethinking, and objective investigators they fancy themselves. Rather, they tend to assimilate what they have been taught and work on solving problems within an accepted framework. Normal science, Kuhn observed, "often suppresses fundamental novelties because they are necessarily subversive of its basic commitments."[6] If some iconoclastic scientists produce results that challenge the prevailing consensus, the research is often dismissed as simply wrong—not as legitimate findings that point to a different sort of paradigm. Eventually, though, the pressure builds to the breaking point and, overnight, the old paradigm collapses and is replaced by a new one.

For example, Ptolemy's theory that the sun revolves around the earth was overthrown by Copernicus's evidence that the planets orbit the sun. When Newton's theories of motion and gravitation could not explain the motion of light, his paradigm gave way to Einstein's theory of relativity. Darwin's theory of natural selection threw out the paradigm of man made in God's image with one scoop of dust. Generally, for this shift to happen, the old scientists have to die off before a revolutionary who has no stake in the system—say, a twenty-six-year-old patent examiner named Albert Einstein—comes along and throws all the cards up in the air. He sweeps away the old paradigm, ushering in a new one. A revolution occurs.

Of course, most mainstream scientists believe we are nowhere close to a Kuhnian-style revolution when it comes to spirituality. They point out that a true revolution requires more than merely trotting out unexplained phenomena and arguing that current explanations should be scrapped. A paradigm shift requires that the new idea make predictions and explain the world in a better way. You can point out that your mother prayed for you and subsequently your fever abated. But before we discard the aspirin, we need a hypothesis about how prayer works and a prediction of what it will cure. For example: reciting the Twenty-third Psalm cures fevers, while the Sermon on the Mount suppresses coughing. Skeptics say that today's "revolutionaries" do not have a theory of "God," much less hard evidence to back it up.

Researchers like neurologist Mario Beauregard at the University of Montreal believe that they are developing such a theory, and that mainstream scientists remain stuck in the "normal science" mode of ignoring the evidence or dismissing it.

"Thomas Kuhn talked of paradigm revolution, and I think we're in the middle of one right now," Beauregard offered. "There are too many data coming from parapsychology, transpersonal psychology, now spiritual neuroscience, quantum physics, and various lines of evidence, all pointing to major failures in the old materialist paradigm. So for me, it's only a matter of time before there will be a major paradigm shift."

He noted that whenever he presents scientific data suggesting we are not biological robots, totally determined by our genes and our neurons—that perhaps we have a spiritual side—the scientists fill the room and clamor to know more. He believes the challenge to the mind-brain problem is gaining followers as more scientists conduct studies on the sly or, more recently, with funding from institutions such as the National Institutes of Health. More important, Beauregard senses a generational shift, the kind that presages the collapse of an old paradigm. The young scientists are restless.

"Many young scientists come to me—not openly, but covertly—and they tell me they greatly admire the kind of work I'm doing," he said. "They don't dare go out publicly yet. But I'm sure it's just a matter of time before we see a true revolution in science because of that. And there will be a big clash, because science has been controlled for centuries by materialists, by atheists."

Clearly, no one knows at this moment whether Mario Beauregard and the other guerrilla scientists are correct. But here's a provocative idea: We will know. New technology that allows us to peer into the brain and record the path of subatomic particles points toward that end. The paradigm will fall or it won't, the spiritual will be acknowledged or marginalized once and for all.

I am reminded of scientist Dean Radin's comment that 96 percent of the universe is "dark matter" or "dark energy"—that is, cosmologists haven't a clue as to what it is. That means that all our scientific theories are built on 4 percent of the observable universe.

"Science is a new enterprise," Radin said. "We are monkeys just out of the trees. And for us to be so arrogant as to imagine we're *close* to understanding the universe is just insane."

How much more arrogant to proclaim we know all about "God." But as the science proceeds, many scientists suspect that the days are numbered for a purely materialist paradigm. They believe that the evidence challenging the matter-only model is building, bolstered by research on meditation, the mechanisms for prayer, and more radical studies on the neurology of near-death experiences. And as the "anomalies" accumulate, these scientists predict the pressure on the dam will grow, and one day the wall of materialism will come crumbling down.

This is the dilemma that Saint Paul voiced in one of his (rare) flashes of intellectual humility.

"Now we see through a glass, darkly," Paul told the Corinthians, "but then face to face: Now I know in part; but then shall I know even as also I am known."

A Personal Paradigm Shift

If I harbor only suspicions about paradigm shifts in science, I am more certain of my own. In the course of looking under the hood of spiritual experience, guided by some smart and very generous scientists, I experienced a series of personal epiphanies about, first, the existence and nature of "God," and second, my own religious beliefs. Let me take these in order.

The question that launched me on this journey could not be more basic: Does God exist? Or, put another way: Is there more than this material reality? This question is at once too ambitious and too modest. Too ambitious because science can never prove that a supernatural being exists; if there is a God, He or She or It operates outside of nature and beyond the reach of scientific measuring instruments. The question is also too modest: I am only reaffirming something that I have always felt to be true. It is difficult to be a Christian, after all, if you do not believe in God or a meaningful universe or an eternal purpose to life. Still, at the outset I worried that I would find a material explanation for every spiritual phenomenon and that my research would drain life of its magic and mystery.

What I have concluded is this: While science cannot *prove* the existence of God, neither can it *disprove* it. In fact, science is entirely consistent with a Being who organized the universe and created life. That may seem an unremarkable conclusion, until you consider that materialists have controlled the levers of science—and with it, a claim on verifiable truth—for centuries.

Materialists can say that this astonishing thing called life (including conscious human beings who are curious about their origins) emerged from a series of random acts, starting from the Big Bang and onward. They can posit that there are 10,500 other universes with no life in them, and we just happened to hit the jackpot by landing in the one universe friendly to life. But isn't it just as plausible—indeed, simpler and more elegant—to postulate that a cosmic Mathematician created the universe to evolve and sustain life?

Or consider spiritual "experience." Materialists can say that our brains evolved so that we would experience profound numinous sensations. For no particular reason, during meditation or an emotional breakdown, on a walk in the mountains or on the cusp of death, some people feel a union with all things, an overpowering love, a light that eviscerates the fear of death. One explanation is that this is a quirk of evolution. But is it not just as plausible to posit that there is a Mind who wired us with the ability to access that Mind and tap into invisible realities?

William James considered the dichotomy between "materialism" and "spiritualism" a hundred years ago. He argued that *both* are internally logical. He asserted that a material worldview that excludes a Creator *and* a spiritual worldview that includes one can both explain natural phenomena, such as the motion of the planets, or the evolution of the universe and life. You can believe either explanations for life and leave it at that. It is when you look toward the *future* that you see how differently the two views play out. As Professor Darren Staloff described it in his lecture on James's pragmatism,[7] materialism holds that eventually our sun will die, Earth will be destroyed, the universe will collapse on itself, and everything we hoped or dreamed or achieved or learned will be for naught. A spiritual worldview, however, leaves room for hope—that all we accomplish, all we are, will be preserved for eternity "if only in the mind of God."

The real distinction between a material and spiritual worldview, James wrote, does not rest "in hair-splitting abstractions about matter's inner essence, or about the metaphysical attributes of God. Materialism means simply the denial that the moral order is eternal, and the cutting off of ultimate hopes; spiritualism means the affirmation of an eternal moral order and *the letting loose of hope.*"[8]

Given the choice, I throw my lot in with hope.

A century after William James wrote those words, I believe there is even more reason to bank on the existence of God. It seems to me that the instruments of brain science are picking up *something* beyond this material world. It seems plausible to me that we perceive the ineffable with spiritual senses, and when the spiritual experience ebbs, it leaves a residue in one's brain or body.

Science is showing that a spiritual experience leaves fingerprints, evidence that a spiritual transaction has occurred. This is hardly a surprising statement: science shows that any significant experience, especially an extraordinary one that lasts longer than thirty minutes, leaves an indelible mark on the brain. A near-death experience after a car crash, a spontaneous mystical experience, hours of prayer and meditation over the years, a seizure in which one hears the "voice of God"—*of course* these mental events sculpt the physical brain that mediates them.

After the encounter, the brain is physiologically different. For adepts in Buddhist meditation, the brain waves sprint along in perfect synchrony. For experts in Christian prayer, the encounter may allow them to quiet certain spatial lobes of their brain so that they feel at one with the universe. A brush with death appears to slow one's brain-wave activity and jump-start a profound spiritual life. Psychedelic drugs and the electrical storm of temporal lobe epilepsy may open the "reducing valve" that filters out information, and allow people to perceive another, nonmaterial, dimension.

At first I was befuddled by these wildly divergent brain patterns and experiences, but then I began to perceive a theme. Simply put, when you bump against the spiritual, something changes. First, your brain begins to operate differently, even at resting state. Second, your interior life is transformed. Your priorities and loves, how you choose to spend your time and with whom you choose to spend it—all that changes in the blink of an eye. And for me at least, the catalyst for this upheaval is God.

Nature and nurture, my DNA and my upbringing, all no doubt inclined me toward embracing this sort of spiritual science. What I did not expect in the course of my research, however, was a radical redefinition of God.

The scientists I interviewed rarely spoke of a personal God, except for geneticist Francis Collins, who described a God who loves math, creates a universe, and hankers for intelligent beings who "wonder if there is more." Usually, they recognized the stunning precision of the laws of the universe, laws that govern particles and galaxies alike. Einstein spoke of a "superior mind," Stephen Hawking of something that "breathes fire into the equations" of the universe. Philosopher Aldous Huxley posited "Mind at Large," Dean Radin suggested an "entangled" tapestry of information, and Larry Dossey spoke of "non-local mind."

These men found God in very large things and very small ones, the vast universe and the microscopic cell. As for me, I set out to test God writ small, but not too small. My subjects were human-sized, and I found evidence of the spiritual painted on the canvas of a person's life. I came to define God by His handiwork: a craftsman who builds the hope of eternity into our genes, a master electrician and chemist who outfits our brains to access another dimension, a guru who rewards our spiritual efforts by allowing us to feel united with all things, an intelligence that pervades every atom and every nanosecond, all time and space, in the throes of death or the ecstasy of life.

This view of God and spiritual reality offers an alternative to superstition. It allows you to steer through an all-or-nothing attitude—that either there is a God who intervenes, depending on His mood and whether you've been naughty or nice; or that "God" is the product of ignorance and we live in a cold, uncaring, random universe. It seems to me that advances in science, and particularly in quantum physics, are offering another description of reality in which all things are guided by and connected to an infinite Mind. This description, of course, echoes the words of mystics down the ages.

As I absorbed the science, I found this was the "God" I could defend most easily—not so much a divine "father" as an infinite creator of law and life. Without realizing it at first, I had looped back to the faith of my childhood. I found myself staring squarely at Mary Baker Eddy's definition of God: "Principle; Mind; Soul; Spirit; Life; Truth; Love; all substance; intelligence."[9] Perhaps Mrs. Eddy was onto something. Perhaps her view of reality presaged what quantum physicists and astrophysicists are finding today.

A Second Look at Christian Science

I remember precisely when I made the connection between my research and Christian Science, and its firm belief that coming in line with spiritual laws will bring about a human solution or healing. I was attending a near-death-experience conference at M. D. Anderson Cancer Center in Houston, listening to Penny Sartori, the spunky intensive-care nurse from Wales. In her presentation, she described the case of Michael Richards, a sixty-year-old patient who suffered from cerebral palsy that left his right hand curled inward and useless. He had been that way since birth. One day while in the hospital, Richards's heart stopped beating, his vital signs suddenly flatlined, and he remained unresponsive for half an hour. He later claimed to have left his body during those thirty minutes and watched the resuscitation from a place somewhere above his bed. He described in precise detail how the doctor had shined a light on his pupils, how Sartori had placed something "long and pink"—a sponge dipped in water—into his mouth to clean it, and how the physiotherapist had repeatedly poked her head around the curtain to see how he was faring.

But there was a twist to Mr. Richards's story that emerged later, as Sartori interviewed him about his near-death experience.

"I said to him, 'While you were in your out-of-body state, was there something you could do that you couldn't do in your body?'" she told the audience. Usually, she explained, patients recalled being able to see without their glasses, or being transported to another location.

"But he misunderstood me and said, 'Oh yeah, look at my hand. I can open my hand.'"

She flashed a photograph onto the screen. The audience gasped. Michael Richards was holding up his right hand for the camera. No longer palsied, his fingers were now straight and he was splaying them like a starfish. A wide grin beamed from his face.

"I discussed this with the physiotherapists, the ones who looked after him, and also independent ones," Sartori told me later. "And they all said this shouldn't be possible, because his hand should be in a permanently contracted position. The tendons were shortened, so in order for him to open his hand, he would have to have an operation to release the tendons. And no such operation was performed. So it's quite remarkable that this happened."

It seemed, at least to the puzzled doctors, that Mr. Richards's contact with death somehow broke the physical laws that held him in bondage. But what if no laws of nature were in fact broken? What if he came in line with spiritual laws that we do not yet understand, in the same way that scientists a millennium ago did not understand the laws of aerodynamics even as they watched birds take flight?

Michael Richards's story persuaded me to open the box I had shelved for more than a decade and reexamine my original faith. It prompted me to reflect on the moment of my most dramatic "healing." I was fourteen years old and suffering from scarlet fever. I remember lying on the bathroom floor, pressing my feverish forehead on the cool tiles. Suddenly, my perspective shifted, and I seemed to be outside my body, viewing it as separate from my conscious identity. (Now I realize it was an out-of-body experience.) I marveled that "I" could be teased apart from that miserable being on the bathroom floor. For a few seconds, I felt no pain and no fever. Then I returned to the prison of my body. But a few minutes later, the fever left me for good and the rash on my face disappeared.

By the end of my year of research about God and science, I was persuaded that these recoveries were not mere coincidence. I was a little closer to understanding the mechanics of how these perplexing healings might work: psychoneuroimmunology, for example, or the power of thought to affect my body, was probably at work. It struck me that these events occur when people who are hurtling along in life suddenly ricochet off of some kind of spiritual truth, and their condition suddenly changes. Not always because of their prayers and theology—Michael Richards was not deep in prayer before his heart stopped; he was chatting with his nurse—but because of, well, location. They walked into grace, like wandering into a patch of sunshine on a gloomy winter day, and were changed by the encounter. Christian Scientists are more deliberate. They don't wander about aimlessly; they set out to find that spiritual law, like a New Yorker who marches out to Broadway to hail a cab. Somehow, it seems, the daily mental practice of Christian Science prayer places a person in the path of spiritual power.

Maybe that is what happened to Michael Richards. Maybe that is what happened to me on the cold tile floor.

Rethinking My Beliefs

When I began my search to understand the spiritual, I was figuratively and literally on a dark mountain in Tennessee, blindly feeling my way along and expecting to stumble with each footfall. I had, by choice, wandered far from the comfortable terrain of the valley, with its paved roads and dinner at eight o'clock sharp. I consciously put my faith to the test. I was frankly worried about the outcome. For the faith I had acquired and polished over the previous decade was a lovely thing. At its center shined a personal God, one who would sit down with me at a wedding reception and share a glass of wine.

As I delved into the science, I realized I need not discard my faith. Rather, I must distinguish it from spiritual experience. A yawning distinction separates the two. Unlike spiritual experience, religious belief can never be tested by a brain scanner or even by historical record. No one can *prove* that Jesus is the Son of God. What religious belief does is attempt to explain in a compelling narrative the unseen reality that lies at the heart of spiritual experience.

I remember once pondering, if God wanted to speak to us, what would He say? For Jews and Christians and Muslims, who believe that the Torah, Bible, and Koran were inspired by God, the answer is simple: God tells stories. He tells stories so we can relate to Him, because each of us is, after all, a story that unfolds minute by minute, day by day. He writes about a hundred-year-old nomad named Abraham who fathered two great tribes. He writes about a shepherd boy named David who felled a giant and became a king. He writes about a prostitute who was forgiven, a blind man who gained his sight, an epileptic boy whose seizures were stilled. For Christians, the greatest story of all tells of a Son who was nailed to a cross, a story of sacrifice and redemption and, most of all, love. And so I fell in love with a story, that of a Christian God who is intensely interested in our lives.

As I studied spirituality, however, it became clear to me that there were elements of my religious beliefs that I would have to discard, or at least view with a skeptical eye. First, I simply cannot read the Bible literally. For one thing, its internal contradictions (witness the two versions of Jesus healing the Centurion's servant in two Gospels) show that the Bible was written by flawed human beings. I believe Scripture is inspired. But when the literal reading of the text conflicts with tested and established science—six-day creation versus evolution, for example—I believe science wins. That does not mean that Scripture is wrong. Any Christian theologian will argue that Genesis is sound theology. But Genesis is not, and never was intended to be, a peer-reviewed article in a scientific journal. Scripture is metaphorical, explaining the world in a way that humans could understand at the time it was written, thousands of years ago. I do not think it was meant to be freeze-framed for all eternity.

That one was easy, since I never did subscribe to the literal inerrancy of the Bible. More problematic was the central tenet of Christianity—that there is only one way to God. When it comes to spiritual experience, there is not one story, and it certainly is not Jesus, not for everyone, not by a long shot. This became blindingly evident as I listened hour after hour to people with radically different story lines—ones that may or may not include God or religion but that involve a transforming encounter with another type of reality that is every bit as powerful to them as the Christian story is to me. These people are as "born again"—with a new outlook, dreams, behaviors, and beliefs—as any evangelical I have met. What story they selected, or what religion they followed, is their own best shot at making sense of something unexplainable—the ache for the eternal, the gnawing suspicion that Earth is not our home.

Embracing a particular faith, I came to believe, is a little like hopping in a car. You can drive wherever you like: some head to Rome, others to Mecca or the Western Wall. Still others exit the highways to find the winding roads called "spiritual but not religious." But what makes the car run is not the color, or the leather upholstery, or even the destination. What makes it run is under the hood—the complex and wondrous mix of physical wiring, chemical reactions, and electrical charges, all of which are the handiwork of a mechanic. Spiritual experience is the engine that transports you from one place to another—and I believe that ability to perceive and engage God is written in each person's genetic code and brain wiring. Religion is the overlay that allows people to navigate the world, and I came to believe that no one religion has an exclusive franchise on God, or truth.

This, I know, puts me outside the circle of much of modern (certainly evangelical) Christianity. I was forced once again to wrestle with the trump card flung in the face of every Christian believer who begins to question that premise: the statement in which Jesus Himself said He was the way, the truth, and the life, and no one comes to God but by Him. I could not reconcile the literal statement with my reporting.

But it seems to me that Jesus' words suggest what we *do*, and not what we *proclaim*. When Jesus says that the way to eternal life is to follow Him, that means trying to live as He did—feeding the poor, helping those who cannot benefit you, loving your enemies, sacrificing rather than promoting yourself, living as if every moment on earth counts for eternity. Can I prove that Jesus is the Son of God? Of course not. Does my instinct tell me that He is the Son of God, and that I should try to emulate Him? It does, and that instinct makes me better.

You do not have to wear a gold cross around your neck to have that kind of heart. Anyone can follow Jesus' example. I have always felt that, but the science persuaded me. Buddhist and Christian meditators reach a very similar spiritual state through the same neurological pathways. The specific content of their beliefs differs, but the underlying spirituality does not. My reporting shows that anyone who has encountered the spiritual—through a brush with death or an unexpected epiphany or even a psychedelic trip—is transformed emotionally and often neurologically. Who am I to start proclaiming one type of experience as real and another as false?

And if I'm wrong? Well, if I'm wrong, I can only hope that the central character in my story, who included a skeptic named Thomas among His closest friends, would see in my questions an honest search for the truth. And, perhaps, He would approve.

In the end, nothing I have learned in the past year undercuts the fundamentals of my Christian faith. I did not lose the young man on the cross, although the old man with a beard can no longer encompass the grandeur and genius of the God I embrace today. I can still believe because my faith rests not on quantum physics or mathematical equations but on an internal witness: a whisper that the story I have chosen is true, that it explains the world. It tells me that the universe is stitched together not just by infinite intelligence but also by love and justice and beauty. These qualities emerge from an elegant Being who revels in things above and beyond the orbit of the planets and the consciousness of man.

My story tells me that we are guided by moral laws, laws that are written in our hearts and genes. It tells me that some behaviors are better than others, some ways to live are more enduring and purposeful. I cannot prove a moral law but I do know instinctively that diving into a rushing river to save a drowning man is nobler than running off to find a rope. There is a hierarchy in morality that is universal—self-sacrifice is everywhere feted and murder is everywhere condemned—but the hierarchy is not derived from physics. It arises from an internal place, a "God-shaped vacuum in the heart of every man," as Blaise Pascal had it, and while that may testify to a moral universe, it is the job of faith to explain the mystery in words that an ordinary person can comprehend. The God who loves math also loves stories, because in the story of faith, the created finds the heart of the Creator.

Fearfully and Wonderfully Made

I end with the question that launched my journey: *Is there more than this?* Yes, I believe there is, and the new science of spirituality buttresses my instinct. Science is showing that you and I are crafted with astonishing precision so that we can, on occasion, peer into a spiritual world and know God. The language of our genes, the chemistry of our bodies, and the wiring of our brains—these are the handiwork of One who longs to be known. And rather than dispel the spiritual, science is cracking it open for all to see.

Of course, these are only my conclusions. But I am not alone in this. This impulse animated the mystics and gave voice to believers down the centuries.

For you formed my inward parts, the Psalmist wrote. *You knitted me together in my mother's womb. I praise you, for I am fearfully and wonderfully made.*

We have all about us the fingerprints of God.

Acknowledgments

It takes a village, it turns out, to write a book. I am astonished at how many people generously offered their help and guidance.

I am indebted to my unsurpassed agent, Rafael Sagalyn, who took a risk on a first-time author with an offbeat topic. Rafe was there at every critical juncture: He stepped in when my book needed to be reshaped and (along with Shannon O'Neill) lent me his flawless editorial guidance. He soothed me when I was upset, and reveled even in small victories. Soup to nuts, Rafe made this book happen.

Jake Morrissey, my editor at Riverhead, believed in this book from the start. He spurred me to write my own story—territory far outside my comfort zone—and on occasion talked me off the ledge. Jake's wit and cheerful intellect made the whole endeavor a pleasure. Thanks, too, to Sarah Bowlin, who graciously answered my tiniest questions.

I owe much to Jeff Goldberg, one of the most talented writers in my generation, and an impossibly nice guy. Jeff *offered* to read my manuscript and spared me from clichés and other embarrassments.

There are many people who contributed to the research, with little or no credit in these pages—but without their efforts, this would be a thin endeavor. My thanks to Kelli Cronin, a resourceful researcher who volunteered her services, and to Mary Glendinning at NPR, who is simply priceless. Rachel Guberman spent countless hours working with me for a pittance; without her, this book would have not been completed before 2020.

I interviewed many scientists, some of whom are featured in the book. Others, however, spent hours explaining the science to me, which was no small task. My huge thanks to Patrick McNamara at Boston

University, who shared years of research with me; Dave Nichols at Purdue University, who explained the neurochemistry; Rick Doblin, who knows all about psychedelic research; geneticist and twins researcher Lindon Eaves at Virginia Commonwealth University; epileptologist Alan Ettinger at North Shore–Long Island Jewish Health System; neurologist Steven Schachter at Harvard University; psychologist Michael McCullough at the University of Miami; Grace Langdon, who walked me through the quantum physics; and Solomon Katz at the University of Pennsylvania and the Metanexus Institute, who is at the vanguard of research into spiritual experience.

Elsevier, the finest scientific publisher in the world, granted me free access to all of its journals for two years; I simply could not have completed the research without its generosity. Mary Ann Liebert Publishers allowed access to its journals as well.

I owe much to my friends at NPR, including Bill Marimow, former Vice President for News (and now editor of *The Philadelphia Inquirer*), who instantly granted me a leave of absence, and Ellen Weiss, current Vice President for News, who graciously held my job open during that year. My thanks as well to Steve Drummond and Cindy Johnston, my NPR editors, who looked the other way when I dragged myself into the office after early-morning bouts of writing.

Nothing can be done without friends and family, and in this, I am rich indeed. One person has inspired me for nearly thirty years. Fred Stocking, my Shakespeare professor at Williams College, told me to "consult my stomach" when I write. He urged me to consider not only whether the story I am telling is accurate, but also whether it rings true—to ensure it reflects not just facts but also my conscience. This has been my goal during my quarter-century of journalism, and in writing this book.

My "small group" has been a bedrock of support, bumping along with me through every valley and helping me negotiate every cliff: Nelda Ackerman, Michelle Brooks, Caroline Comport, Kelly Cowles, Cherie

Harder, Jody Hassett Sanchez, Shawn Walters. I am fortunate to count you as my dearest friends.

As for my family, what can I say about Mom, who listened breathlessly to the accounts of my findings every single day? You inspired me and questioned me, and your genuine interest encouraged me to believe that perhaps a few people would read my book. My dad, Gene Bradley, and his wonderful wife, Nancy, never ceased to be excited about this project. To my sister-in-law, Katherine, who loved the idea from the start, read through the early, dismal drafts, and offered spot-on editorial guidance, thank you. To my brother, David, who has always seen more ability in me than I see in myself, you will always be my *kuya.* And thanks to the world's greatest nephews, Spencer, Carter, and Adam, who listened to my exploits around the dinner table.

I am blessed with an extraordinary stepdaughter, the beautiful and brilliant Vivian Grace Hagerty, who gives me great joy and who made some sage comments on the manuscript—not bad for a then thirteen-year-old. Surely the person to whom I am most indebted is my husband, Devin, who told me to take risks, to write that proposal, to take that leave of absence, to blaze through the research and writing in little more than a year. You comforted me when I cried after that little peyote episode, laughed at my "God helmet" stories, and edited my manuscript line by line. What can I say? You are a rock star.

Notes

CHAPTER 2. THE GOD WHO BREAKS AND ENTERS

1. Sophy Burnham, *The Ecstatic Journey* (New York: Ballantine, 1997).
2. Here is Saint Paul's description to the Corinthians: "I will go on to visions and revelations from the Lord. I know a man in Christ who fourteen years ago was caught up to the third heaven. Whether it was in the body or out of the body I do not know—God knows. And I know that this man—whether in the body or apart from the body I do not know, but God knows—was caught up to paradise. He heard inexpressible things, things that man is not permitted to tell" (2 Corinthians 12:1–4).
3. *The Revelations of Divine Love in Sixteen Showings Made to Dame Julian of Norwich*, trans. M. L. Del Mastro (St. Louis: Liguori Publications, 1994), chapter 27.
4. William James, *The Varieties of Religious Experience: A Study in Human Nature* (Cambridge, Mass.: Harvard University Press, 1985; originally published 1902), p. 3.
5. Ibid., p. 138.
6. Ibid., p. 124.
7. Ibid., p. 415.
8. Ibid., p. 422 (italics mine).
9. Ibid., p. 461.
10. J. H. Leuba, *The Psychology of Religious Mysticism* (New York: Harcourt, Brace, 1925). For an excellent summary of spirituality and science, see B. Spilka et al., eds., *The Psychology of Religion: An Empirical Approach* (New York: Guilford Press, 2003), pp. 291–98.
11. Sigmund Freud, *Civilization and Its Discontents*, trans. J. Strachey (New York: W. W. Norton, 1961; originally published 1930).
12. Émile Durkheim, *The Elementary Forms of the Religious Life: A Study in Religious Sociology*, trans. J. W. Swain (London: Allen & Unwin, 1915).
13. R. M. Bucke, *Cosmic Consciousness: A Study of the Evolution of the Human Mind* (Hyde Park, N.Y.: University Books, 1961; originally published 1901).
14. C. G. Jung, *Archetypes of the Collective Unconscious*, in H. Read, M. Fordham, and G. Adler, eds., *The Collected Works of C. G. Jung*, trans. R. F. C. Hull, 2nd ed. (Princeton, N.J.: Princeton University Press, 1968; originally published 1954), vol. 9, part 1, pp. 3–41.
15. A. H. Maslow, *Religions, Values, and Peak Experiences* (Columbus: Ohio State University Press, 1964).
16. John B. Watson, "Psychology As the Behaviorists View It," *Psychological Review* 20 (1913): 158–77.
17. The polls were reported in *Nature*, April 3, 1997, 435–36, and July 23, 1998, 313.
18. W. Miller, *Quantum Change* (New York: Guilford Press, 2001).
19. Ibid., p. 83.
20. Ibid., p. 85.
21. I heard a lot about orgasm in my research. Sophy Burnham told me: "I'd wake up, just with radiance running through my body. It was as if I'd been made love to by God." Llewellyn Vaughan-Lee, a Sufi mystic I interviewed, was more explicit. "The Sufis would say that the

external relationship between two people is a pale reflection of the relationship between lover and beloved," he explained. As I listened to Llewellyn, it struck me that his words and his faraway tone reflected something, well, erotic. "If you really practice mediation, and you're taken along that path, you experience a much more enduring bliss. It's very sexual but it's beyond sexual. The orgasm doesn't happen in the sex tracts, but happens in the heart. And that is unbelievable. You have experiences of cosmic love and cosmic bliss. And rather than lasting for a moment, it can last for hours. It is a quite wonderful experience." I was dumbfounded. "How many people can achieve that?" I blurted. Llewellyn burst out laughing. "It's what mystics long for, that ecstatic union with God," he said, not really answering my question, but I suppose I did not expect him to give me numbers. "It is very, very erotic, but it is much deeper than that. It's what lovers long for." Of course, the mystics of old alluded to this. Consider Teresa of Ávila, whose erotic notions of God were immortalized in a statue by Bernini. In one of her mystical states, she wrote, she encountered an angel; he was not tall, but very beautiful, and his face was "aflame." "In his hands I saw a long golden spear and at the end of the iron tip I seemed to see a point of fire. With this he seemed to pierce my heart several times so that it penetrated to my entrails. When he drew it out, I thought he was drawing them out with it and he left me completely afire with a great love for God. The pain was so sharp that it made me utter several moans; and so excessive was the sweetness caused me by the intense pain that one can never wish to lose it, nor will one's soul be content with anything less than God." Teresa of Ávila, *The Life of Saint Teresa by Herself*, trans. J. M. Cohen (New York: Penguin, 1988), chapter 29. Why, I thought, would a spiritual experience give a thousand-watt jolt to the body? The answer is simple—and, happily, it opens the door for scientists to explore. We are, after all, physical beings. If there is such a thing as a veiled reality, how else are we to experience it except physically, with our synapses firing and our brains activated and our hearts racing? Given our makeup, is it really surprising that spiritual virtuosos might enjoy a little afternoon delight?

22. Miller, *Quantum Change*, p. 106.

23. National Opinion Research Center, University of Chicago, "Spiritual and Religious Transformation in America: The National Spiritual Transformation Study" (report prepared for Metanexus Institute, Philadelphia, June 2005). To find out just how present "God" is for the average American, I called up Tom W. Smith at the National Opinion Research Center. In June 2005, he and his team of researchers completed their twenty-fifth General Social Survey. They had interviewed 1,300 people at length about their spiritual journeys. For most in the survey, that spiritual experience was relatively tame: being "born again" in a Baptist church, feeling moved by a sermon, or inspired by a hymn when singing in the church choir. It could be an "aha" moment, often in the wake of a death or a tragedy, when people turn to God. But some described less run-of-the-mill events. "For some people, angels played the trumpet," Smith observed with admirable restraint. "We got every standard litany of change: God talking to them, floating, out-of-body experiences, near-death experiences, tunnels with light." According to Smith's research, 18 percent of Americans reported experiences that could be listed on the pages of the *Diagnostic and Statistical Manual of Mental Disorders*; hearing God talking to them, floating outside their bodies, dying yet remaining conscious, contacting or being contacted by the dead, feeling a supernatural "jolt" and seeing a light (in a tunnel or otherwise), seeing a spirit, having a physical sensation of God. Are all these people crazy? Am I crazy, since I have experienced some of those phenomena myself? A century ago, these people might have been candidates for the asylum, or lobotomy, or both. But in recent decades, studies have found that people who experience mystical states are quite stable. They are better educated than the average American, wealthier, and relatively mature (in their forties and fifties). Psychological testing reveals that they tend to be open to new experiences, have a breadth of interests, and are innovative, tolerant of ambiguity, and creative. True, they

are more easily hypnotized and prone toward fantasy. But when you pass a mystic walking down the street, she probably won't be muttering obscenities. Chances are she's healthy and smiling. As pollster, author, and priest Andrew Greeley put it, "Mystics are happier. Ecstasy is good for you." See M. A. Thalbourne, "A Note on the Greeley Measure of Mystical Experience," *International Journal for the Psychology of Religion*, 14(3): 215–22.

24. He asked that I not use his real name, to protect his reputation.

25. Naturally, William James recognized this phenomenon a century ago. Mystical states, he wrote in *The Varieties of Religious Experience*, allow the mystic to become one with the Absolute, and be aware of that oneness—a tradition that defied "clime or creed." "In Hinduism, in Neoplatonism, in Sufism, in Christian mysticism, in Whitmanism, we find the same recurring note, so that there is about mystical utterances an eternal unanimity which ought to make a critic stop and think, and which brings it about that the mystical classics have, as has been said, neither birthday nor native land." James, *Varieties*, p. 324.

26. John 14:6.

CHAPTER 3. THE BIOLOGY OF BELIEF

1. Norman Cousins also took massive doses of vitamin C. But his case seemed to demonstrate that positive emotions are good for your health. See his *Anatomy of an Illness as Perceived by the Patient: Reflections on Healing and Regeneration* (New York: W. W. Norton, 1979).

2. A study conducted in fifty-two countries found that psychosocial stress accounted for about 40 percent of heart attacks. Salim Yusuf et al., "Effect of Potentially Modifiable Risk Factors Associated with Myocardial Infarction in 52 Countries (The INTERHEART Study): Case-Control Study," *The Lancet* 364, no. 9348 (September 11–17, 2004): 937–52.

3. Janice K. Kiecolt-Glaser and her colleagues found that chronic stress altered the immune response to a flu virus vaccine in older adults. They looked at the responses of "caregivers" (who took care of spouses with Alzheimer's disease for at least three years) and compared them with a control group of less stressed people. In the control group, 70 percent responded to the flu vaccine, contrasted with 35 percent of caregivers. This suggests that stress reduced the number of people who produced the protective antibody to the virus by 50 percent. Janice K. Kiecolt-Glaser et al., "Chronic Stress Alters the Immune Response to Influenza Virus Vaccine in Older Adults," *Proceedings of the National Academy of Sciences* 93 (1996): 3043–47.

4. J. K. Kiecolt-Glaser and R. Glaser, "Psychoneuroimmunology and Cancer: Fact or Fiction?" *European Journal of Cancer* 35 (1999): 1603–7.

5. In one study, British researchers followed the cases of 578 women diagnosed with early-stage breast cancer. Five years after diagnosis, those women who scored high on an anxiety and depression scale had a significantly increased risk of death. Those scoring high on the helplessness/hopelessness scale had a higher risk of relapse and death M. Watson et al., "Influence of Psychological Response on Survival in Breast Cancer: A Population-Base Cohort Study," *The Lancet* 354 (1999): 1331–36. In another study, researchers followed sixty-two breast cancer patients over five, ten, and fifteen years. They found that women who responded with a "fighting spirit" or with denial (what the researchers called "positive avoidance") were significantly more likely to be alive and well for at least fifteen years. By contrast, women with fatalistic or helpless outlooks fared far more poorly. S. Greer et al., "Psychological Response to Breast Cancer and 15-Year Outcome," *The Lancet* 335 (1990): 49–50.

6. J. Kabat-Zinn and his colleagues studied psoriasis, which is an uncontrolled cell proliferation of a layer of the skin and can cover the whole body. Stress makes it worse. The researchers found that those who consciously reduced their stress levels through meditation healed more quickly than those who used only mainstream treatments. For those who

meditated, the condition cleared in 100 days, compared with 125 days for those who did not meditate. J. Kabat-Zinn et al., "Influence of a Mindfulness Meditation-Based Stress Reduction Intervention on Rates of Skin Clearing in Patients with Moderate to Severe Psoriasis Undergoing Phototherapy (UVB) and Photochemotherapy (PUVA)," *Psychosomatic Medicine* 60 (1998): 625–32.

7. G. Ironson, R. Stuetzle, and M. A. Fletcher, "An Increase in Religiousness/Spirituality Occurs After HIV Diagnosis and Predicts Slower Disease Progression Over 4 Years in People with HIV," *Journal of General Internal Medicine* 21 (supplement; 2006): S62–68.

8. G. Ironson et al., "View of God Is Associated with Disease Progression in HIV." Paper presented at the annual meeting of the Society of Behavioral Medicine, March 22–25, 2006, San Francisco. Abstract published in *Annals of Behavioral Medicine* 31 (supplement): S074.

9. G. Ironson at al., "The Ironson-Woods Spirituality/Religiousness Index Is Associated with Long Survival, Health Behaviors, Less Stress, and Low Cortisol in People with HIV/AIDS," *Annals of Behavioral Medicine* 24, no. 1: 34–38 (special issue on religion and health).

10. R. C. Byrd, "Positive Therapeutic Effects of Intercessory Prayer in a Coronary Care Unit," *Southern Medical Journal* 81 (1988): 826–29.

11. F. Sicher and colleagues followed forty patients with advanced AIDS. Half received different types of prayer and psychic healing for ten weeks; the other half did not. The patients receiving prayer developed fewer AIDS-defining illnesses, experienced less severe illness when they were sick, made fewer trips to the hospital or their doctors, and spent less time in the hospital. However, there were no significant differences in CD4 cell counts, a biological measure that charts the progression of the disease. F. Sicher et al., "A Randomized, Double-Blind Study of the Effects of Distant Healing in a Population with Advanced AIDS," *Western Journal of Medicine* 169, no. 6 (1998): 356–63.

12. W. S. Harris studied 990 patients admitted to a coronary care unit at a private hospital. Prayer intercessors were each given the first name of one patient (and it was guaranteed that each did not know the patient) and prayed for that patient every day for four weeks. Half of the patients received no prayer. The group receiving prayer did better overall. W. S. Harris et al., "A Randomized, Controlled Trial of the Effects of Remote Intercessory Prayer on Outcomes in Patients Admitted to the Coronary Care Unit," *Archives of Internal Medicine* 159 (1999): 22–78.

13. The 219 women were in Seoul, while the prayer groups lived in the United States, Canada, and Australia. K. Y. Cha, D. P. Wirth, and R. A. Lobo, "Does Prayer Influence the Success of In Vitro Fertilization-Embryo Transfer? Report of a Masked, Randomized Trial," *Journal of Reproductive Medicine* 46 (2001): 781–87. Later, one of the researchers was found guilty of fraud in an unrelated study, which cast doubt on these findings in the minds of many researchers.

14. I love this study. Twenty-two bush babies (*Otolemur garnettii*) with "chronically self-injurious behavior" were monitored over four weeks. Half received prayer and medication on their wounds each day; the other half received only medication. The bush babies receiving prayer healed more quickly, for both biological reasons (a greater increase in red blood cells, for example) and behavioral reasons (they didn't lick their wounds as much, which allowed them to heal). K. T. Lesniak, "The Effect of Intercessory Prayer on Wound Healing in Nonhuman Primates," *Alternative Therapies in Health & Medicine* 12 (2006): 42–48.

15. L. Leibovici, "Effects of Remote, Retroactive Intercessory Prayer on Outcomes in Patients with Bloodstream Infection: Randomized, Controlled Trial," *British Medical Journal* 323: 1450–51.

16. J. M. Aviles and colleagues monitored 799 coronary care unit patients between 1997 and 1999. Half were prayed for by five different intercessors once a week for twenty-six weeks. At the end, the group receiving prayer scored slightly better (but not in a statistically significant measure)

in areas such as death, cardiac arrest, rehospitalization for cardiovascular disease, coronary revascularization, and emergency department visits for cardiovascular disease. J. M. Aviles et al., "Intercessory Prayer and Cardiovascular Disease Progression in a Coronary Care Unit Population: A Randomized, Controlled Trial," *Mayo Clinic Proceedings* 76 (2001): 1192–98.

17. The researchers measured not just prayer among the 748 patients but also an alternative therapy of music, imagery, and touch. Neither prayer nor the alternative therapies seemed to affect the outcomes as measured by death or major cardiovascular events. M. W. Krukoff et al., "Music, Imagery, Touch, and Prayer as Adjuncts to Interventional Cardiac Care: The Monitoring and Actualisation of Noetic Trainings (MANTRA) II Randomized Study," *The Lancet* 366 (2005): 211–17.

18. Researchers at California Pacific Medical Center split 156 patients into roughly three categories: those who received prayer or distant healing from professional healers for ten weeks; those who received prayer or distant healing from nurses who had no prior training in healing (also ten weeks); and those who received nothing. No significant effects were observed for those who received prayer from trained healers or nurses. J. A. Astin et al., "The Efficacy of Distant Healing for Human Immunodeficiency Virus: Results of a Randomized Trial," *Alternative Therapies in Health & Medicine* 12 (2006): 36–41. However, there was what I consider a fatal flaw in this study, something the authors called a "limitation." Namely, they lost much of their data: 40 percent of the prayer groups and 24 percent of the control group never showed up at the end of the ten-week period to be analyzed. This made me wonder about the robust nature of other studies. And it prompted Dr. Larry Dossey, who sent me the article, to note: "I don't know why people publish stuff like this. It just pollutes the literature. Now people will cite this study as evidence that prayer is worthless in HIV/AIDS, a wholly unjustified conclusion based on this experiment."

19. Researchers studied forty (mostly female) patients with rheumatoid arthritis. Some received in-person intercessory prayer; others received prayer from people far away as well as in person. Patients who received prayer in person were found to improve significantly, but the distant prayer did not add any benefit. D. A. Matthews, S. M. Marlowe, and F. S. MacNutt, "Effects of Intercessory Prayer on Patients with Rheumatoid Arthritis," *Southern Medical Journal* 93 (2000): 1177–86.

20. The study looked at ninety-five patients with end-stage renal disease. The upshot was that the people who expected to be prayed for *said* they felt significantly better than did those who expected to receive another mental treatment (positive visualization). But on every other measure, prayer made no difference. W. J. Matthew, J. M. Conti, and S. G. Sireci, "The Effects of Intercessory Prayer, Positive Visualization, and Expectancy on the Well-being of Kidney Dialysis Patients," *Alternative Therapies in Health & Medicine* 7 (2001): 42–52.

21. E. Harkness, N. Abbot, and E. Ernst, "A Randomized Trial of Distant Healing for Skin Warts," *American Journal of Medicine* 10 (2000): 448–52.

22. H. Benson et al., "Study of the Therapeutic Effects of Intercessory Prayer (STEP) in Cardiac Bypass Patients: A Multicenter Randomized Trial of Uncertainty and Certainty of Receiving Intercessory Prayer," *American Heart Journal* 151 (2006): 934–42.

23. Richard P. Sloan, *Blind Faith: The Unholy Alliance of Religion and Medicine* (New York: St. Martin's Press, 2008).

24. One of the more remarkable involved my own infant ears. When I was a few months old, I developed an excruciating ear infection. I shrieked for several days straight to announce the problem. After days of intense prayer by my mother and our Christian Science practitioner, Mrs. Wooden, I grew quiet. Gratified by the silence, my mother tiptoed into my room. Lying on either side of my head, next to my ears, were two pieces of a hard yellow substance in the shape of honeycombs—double mastoids that had somehow emerged from my ears on their

own. Of course, a skeptic would say that my mother did not document this "healing" by getting a doctor to confirm it, or that if such a "healing" did occur, you could say it was the natural course of events. Everything depends on how you read the "evidence."

25. See Astin et al., "The Efficacy of Distant Healing."

CHAPTER 4. THE TRIGGERS FOR GOD

1. Granqvist found that people who experience sudden religious conversions more often have "insecure attachment histories" (distant relationship with parents, creating a need to compensate, which prompted them to see God as surrogate parent). These people are nearly twice as likely to have sudden conversions as do people who have secure relationships with their parents. They tend to be more religious if their parents are less so, and vice versa (a sort of "I'll show you" phenomenon). People who have secure relationships with their parents tend to experience gradual conversions and religious changes. These children develop a similar relationship with "God" as their parents did, and also tend to adopt their parents' religious or nonreligious standards. P. Granqvist and L. Kirkpatrick, "Religious Conversion and Perceived Childhood Attachment: A Meta-analysis," *International Journal for the Psychology of Religion* 14 (2004): 223–50.

2. Jerome Kagan, a Harvard child psychologist, told me in an interview that he's found a link between religiousness and children who are "high reactive." He has been following five hundred white middle-class adolescents (now sixteen years old) since they were four months old. Infants who were high-reactive—that is, they squirmed and waved their legs and arms at the smallest stimulus, such as a mobile over their cribs—tended to show greater cortical arousal throughout the years. They were more anxious and tense. By the time the children reached adolescence, twice as many of those who were high-reactive as infants had become religious, as compared with the low-reactive kids. Kagan theorized that the children were using religiousness as a coping mechanism to help them reduce tension. (He notes that the adolescents who were high-reactive and *not* religious—all three of them—were in therapy and on drugs.) "A spiritual outlook is helpful," Kagan told me, "because it says, 'Things are going to be okay, you're in good hands, there is a supernatural force, and this supernatural force will take care of you. You just be good, be kind to others, believe in some sort of supernatural force.'"

3. B. Zinnbauer and K. Pargament, "Spiritual Conversion: A Study of Religious Change Among College Students," *Journal for the Scientific Study of Religion* 37 (1998): 161–80. Since the 1800s, studies have shown that conversion is a radical change following a period of stress (often crisis). Some have found that 80 percent of converts reported serious distress, including feelings of despair, doubts about self-worth, fear of rejection, estrangement. Others found that converts had problematic relationships with their fathers, and that they were actively seeking a conversion experience to resolve life difficulties.

Zinnbauer and Pargament studied 130 college students at a Christian college, ages eighteen to twenty-eight. They found that religious converts—those who convert to a sacred, transcendent force such as Jesus, God, or Allah, and feel connected to that force—have undergone more stress in the six months before their conversion. They did not actually have more stressful lives than the nonconverts, but they thought they did. They also had a greater sense of personal inadequacy and limitation. One problem with this study, of course, is that students attending a Christian school do not greatly differ from one another, so you have the sense that the researchers are straining at gnats.

4. P. McNamara, ed., *Where God and Science Meet: How Brain and Evolutionary Studies Alter Our Understanding of Religion*, 3 vols. (Westport, Conn., and London: Praeger Perspectives, 2006).

5. Raymond F. Paloutzian, Erica L. Swenson, and Patrick McNamara, "Religious Conversion, Spiritual Transformation, and the Neurocognition of Meaning Making," in McNamara, *Where God and Science Meet*, vol. 2: *The Neurology of Religious Experience*, pp. 151–69.

CHAPTER 5. HUNTING FOR THE GOD GENE

1. W. Miller, *Quantum Change* (New York: Guilford Press, 2001), p. 94.
2. The Gospel of Thomas is one of the Gnostic gospels, not included in the Bible. Don Eaton closely paraphrased verse 77, in which Jesus tells his disciples: "Split a piece of wood; I am there. Lift up the stone, and you will find me there."
3. C. R. Cloninger, D. M. Svrakic, and T. R. Przybeck, "A Psychobiological Model for Temperament and Character," *Archives of General Psychiatry* 50 (1993): 975–90. Cloninger's self-transcendence trait is determined by three criteria. One is called "spiritual acceptance versus rational materialism" and involves phenomena such as mystical experience or a belief in miracles, the supernatural, and a force greater than oneself directing one's life. Another is "transpersonal identification," that is, a connectedness to the universe and everything in it, including nature and people. Finally, there is "self-forgetfulness," or absorption with beauty, music, and the task at hand, to the point of forgetting oneself, time, and space. One's spirituality is measured by how one responds, "True" or "False," to a series of statements such as "I believe in miracles" or "Sometimes I have felt like I was part of something with no limits or boundaries in time or space."
4. Researchers make a distinction between spirituality, which involves personal experience, and religiosity, which pertains to doctrinal beliefs and external religious practices. One way to distinguish the two is to consider intrinsic and extrinsic religiousness. Spirituality is often equated with intrinsic religiousness—that is, an inward-looking faith that is not necessarily tethered to a particular religion; it involves private prayer, meditation, and a strong sense of God's presence; one's whole approach to life is based on religion. Extrinsic religion is outward-looking: I go to my church or synagogue to spend time with my friends; I pray because I've been taught to pray; I don't let religion affect my daily life. Saroglou found that people who are intrinsically religious (more spiritual than traditional) score high on agreeableness and conscientiousness, and openness to experience. Those who are more extrinsically religious (traditional) are conscientious and agreeable, but are not open to new experience. They also show high levels of neuroticism. V. Saroglou, "Religion and the Five Factors of Personality: A Meta-analytic Review," *Personality and Individual Differences* 32 (2001): 15–25.
5. On the heritability of spirituality, see the following:

K. Kirk, L. Eaves, and M. Martin, "Self-Transcendence as a Measure of Spirituality in a Sample of Older Australian Twins," *Twin Research* 2 (1999): 81–87. In a sample of more than 3,000 twins (identical and fraternal), genes appeared to explain 41 percent of the variation of spirituality in women, and 37 percent in men.

L. Eaves, B. D'Onofrio, and R. Russell, "Transmission of Religion and Attitudes," *Twin Research* 2 (1999): 59–61. The researchers found that variation in personality is partly genetic, but there are large effects from the family environment.

T. Bouchard et al., "Intrinsic and Extrinsic Religiousness: Genetic and Environmental Influences and Personality Correlates," *Twin Research* 2 (1999): 88–98. This study of thirty-five identical and thirty-seven fraternal twin pairs raised apart found that intrinsic religiousness is 43 percent heritable; extrinsic religion is 39 percent heritable. The rest is attributed to nonshared environment.

6. Dean Hamer, *The God Gene* (New York: Doubleday, 2004).
7. Hamer's idea of a "gay gene," as proposed in his 1995 book *The Science of Desire* (coauthored with Peter Copeland), stirred controversy, and large sales; the book was a *New York Times*

Notable Book of the Year. Hamer's findings never appeared in a peer-reviewed scientific journal, and scientists were unable to replicate them. By the time other scientists were commenting that they could not find a gene that inclines one toward homosexuality, Hamer had moved on to his next project: the God gene.

8. Francis Collins, *The Language of God* (New York: Free Press, 2006).

9. This has put Collins in hot water with evangelicals, who take the Bible literally. For example, he believes in evolution, and when he said so in *The Language of God*, major evangelical figures refused to endorse his book.

10. D. Comings et al., "The DRD4 Gene and the Spiritual Transcendence Scale of the Character Temperament Index," *Psychiatric Genetics* 10 (2000): 185–89. (Note that neither the serotonin study nor the dopamine study has been replicated.)

11. As in many studies, the researchers tried to compare similar subjects, and thus did not compare across gender. In addition, most of this spirituality research is done on a shoestring, and researchers often recruit their subjects from already existing treatment programs or trials; therefore, researchers do not have much of a choice of subjects.

12. In both the serotonin and dopamine studies, it was the subset "spiritual acceptance versus material rationalism" that accounted for most of the difference in spirituality. This makes intuitive sense to me. It seems logical that believing in miracles or that one's life is directed by a spiritual force greater than any human being, or feeling in contact with a divine being, describes a more classic spirituality. Transpersonal identification (feeling connected to others and being willing to sacrifice for the good of other people, animals, nature, and the world) seems more a qualification for the Humane Society than evidence of a spiritual worldview. Self-forgetfulness (losing oneself in thought, time, and space) appears to me a measure of one's ability to focus—a trait surely found as readily in the atheist biologist or agnostic violinist as in the pastor or mystic.

13. D. E. Comings et al., "A Multivariate Analysis of 59 Candidate Genes in Personality Traits: The Temperament and Character Inventory," *Clinical Genetics* 58 (2000): 376.

14. Comings et al., "The DRD4 Gene," p. 188.

15. J. Borg et al., "The Serotonin System and Spiritual Experiences," *American Journal of Psychiatry* 160 (2003): 1965–69.

16. What they found was an inverse relationship: the higher the subjects' spirituality score, the fewer the number of serotonin receptors that lit up. A couple of theories were offered to explain what that might mean. One was that spiritual people have less of that neurotransmitter. The researchers reasoned that the serotonin system regulates a person's perception and the various sights, sounds, and other stimuli that reach his awareness; therefore, if there is a weak "sensory filter," they wrote, that would allow for "increased perception and decreased inhibition." In other words, the filter is more loosely woven, so that more spiritual experiences get through. Think about a soccer team playing without a goalie: it's much easier for "God" to kick spiritual feelings and ideas into one's psyche if a key player in the defensive line is missing.

The researchers considered another theory. Maybe the serotonin neurons were firing a lot, and so when the radioactive tracer drug came along looking for a place to dock, the receptors were otherwise occupied. Imagine you are throwing your six-year-old a birthday party in your backyard. His friends begin to arrive, and you launch the festivities with a game of musical chairs. More and more six-year-olds arrive, and you run out of chairs. Soon, too soon, you have two dozen six-year-olds cavorting around the yard, kicking up the grass and trampling the daffodils. Such euphoria, such joy and awe! That is a transcendent moment in the serotonin system.

17. Borg et al., "The Serotonin System," 1965.

CHAPTER 6. ISN'T GOD A TRIP?

1. J. H. Halpern et al., "Psychological and Cognitive Effects of Long-Term Peyote Use Among Native Americans," *Biological Psychiatry* 58 (2005): 624–31.

2. R. R. Griffiths et al., "Psilocybin Can Occasion Mystical-Type Experiences Having Substantial and Sustained Personal Meaning and Spiritual Significance," *Journal of Psychopharmacology* 187 (2006): 268–83.

3. For the follow-up study, see R. R. Griffiths et al., "Mystical-Type Experiences Occasioned by Psilocybin Mediate the Attribution of Personal Meaning and Spiritual Significance 14 Months Later," *Journal of Psychopharmacology*, published online July 2008.

The Johns Hopkins study was modeled after the "Good Friday experiment" of 1962, the most famous study to attempt to plumb spiritual experience through psychedelics. On Good Friday, 1962, twenty divinity students gathered in the basement of Boston University's Marsh Chapel. Half of them received a capsule (30 mg) of psilocybin, the active ingredient in psychedelic mushrooms. The other half received a placebo. After eight hours of exploring alternative states of consciousness, the psilocybin group returned to consensus reality. Four of the ten men who had received the psilocybin had encountered a full-blown mystical experience, including transcendence of time and space, and unity with all things, a sense of sacredness. Four others experienced most of the characteristics of mystical experience. None of the control group enjoyed much of anything except profound boredom. See W. Pahnke, "Drugs and Mysticism: An Analysis of the Relationship Between Psychedelic Drugs and the Mystical Consciousness" (Ph.D. dissertation in Religion and Society, Harvard University Graduate School of Arts and Sciences, 1963).

Griffiths used some of the same measures of mystical experience that Walter Pahnke employed in his "mystical consciousness" experiments. M. W. Johnson, W. A. Richards, and R. R. Griffiths, "Human Hallucinogen Research: Guidelines for Safety," *Journal of Psychopharmacology,* published online July 2008.

4. That serotonin receptor gene made an appearance in the previous chapter. Swedish researchers found that serotonin HT1A—which acts as a docking station to allow the chemical to enter—seemed to affect whether their subjects scored high in spirituality, or "self-transcendence."

5. Andrew Newberg at the University of Pennsylvania has found the opposite behavior in the parietal lobes. A study of Buddhist monks deep in meditation and Franciscan nuns deep in prayer showed that the parietal lobes grew quiescent, essentially eliminating the boundaries between the subjects' perception of themselves and the rest of the universe. This is discussed in detail in chapter 8. The research is still in its infancy: the cartographers of the brain are just beginning their work. It will take many more scouting trips, many more studies, to plot the neural landscape successfully, especially as it's not only a physical landscape but a spiritual one as well.

6. David Nichols subscribes to another theory about chemical heaven or hell. A pharmacologist at Purdue University who has studied the chemistry of mystical experiences, he speculates that when you take enough drugs, the front part of your brain that tries to make sense of the world is working overtime. But it doesn't have much real stuff—real sights or sounds—to work with. That could be because the thalamus, the gateway that lets in sensory information, shuts down. Or it could be because you are closing your eyes and withdrawing into your own little world. "What the brain is actually processing," Nichols posits, "is not sensory information, but is related to subconscious things: your dreams, desires, fears, anxieties, memories. You're still conscious, and your brain has to process something, so it processes all your intuitions and feelings and imagination and fantasies. Now you're into this realm where I would say the mystical experience occurs." In other words, the brain is creating

its own reality—its own heaven and hell—because it does not have external sensory information to work from.

7. Other researchers, who have different theories about the chemical process of spiritual experience, point out that the dosage used in Vollenweider's subjects was only enough to disrupt the senses, not plunge a person into a full-blown hallucinogenic experience. The dosages of psilocybin in the Good Friday experiment and in Roland Griffiths's study at Johns Hopkins were much higher.

8. Aldous Huxley, *The Doors of Perception* (New York: Perennial Classics, 2004; originally published 1954), p. 23.

9. Ibid., p. 26.

10. E. C. Kast and V. J. Collins, "Lysergic Acid Diethylamide as an Analgesic Agent," *Anesthesia & Analgesia* 43 (1964): 285–91.

11. See D. E. Nichols, "Commentary on: 'Psilocybin can occasion mystical-type experiences having substantial and sustained personal meaning and spiritual significance by Griffiths et al.,'" *Journal of Psychopharmacology* 187 (2006): 284–86.

12. Stanislav Grof, *The Ultimate Journey: Consciousness and the Mystery of Death* (Ben Lomond, Calif.: Multidisciplinary Association for Psychedelic Studies, 2006). One patient was Jesse, an unmarried thirty-two-year-old man with tumorous masses on his face and neck. A strict Catholic, Jesse had been divorced for years and was terrified of dying; he felt certain he was headed for hell or for a void of nothingness. Grof administered 90 milligrams of DMT, and after a traumatic start to his trip, Jesse saw a gigantic ball of fire. "He experienced a Last Judgment scene where God [Jehovah] was weighing his good and evil deeds," Grof reported later. "The positive aspects of his life were found to outweigh his sins and transgressions. Jesse felt as if a prison had opened up and he had been set free. At this point he heard sounds of celestial music and angelic singing, and he began to understand the meaning of his experience. A profound message came to him through some supernatural, nonverbal channels and permeated his whole being: 'When you die, your body will be destroyed, but you will be saved; your soul will be with you all the time. You will come back to earth, you will be living again, but you do not know what you will be on the next earth.'" Grof reported that Jesse emerged from the experience believing in reincarnation; and with no "end" in sight, his anxiety and depression lifted. He died five days later. Grof wrote, "It almost seemed as if he were hurrying to get a new body on the 'next earth.'"

Another patient, Ted, was a twenty-six-year-old African-American man with a wife and three children, who was suffering from inoperable colon cancer. Ted received 300 micrograms of LSD—a hefty dose—on three separate occasions. During the first, the Vietnam vet had visions of war scenes and children dying of epidemic diseases, followed by an ecstatic assurance that no one actually dies. His pain levels dropped so dramatically after the session that he soon took on a volunteer job. Grof reported that in the second session, Ted experienced his own death, "during which God appeared to him as a brilliant source of light. This was a very beautiful and comforting episode, as God told him there was nothing to fear and assured him that everything would be all right." For me, the most arresting incident occurred later, when surgeons were making a last, desperate attempt to save Ted's life. During an operation, Ted suffered two cardiac arrests resulting in clinical death. He was resuscitated both times. Later he told Grof that the transition to the afterlife was familiar territory for him, because he had traveled there before, during his LSD sessions. "Without the LSD sessions, I would have been scared by what was happening," he told Grof. "But knowing these states, I was not afraid at all."

13. Albert A. Kurland et al., "Psychedelic Drug Assisted Psychotherapy in Patients with Terminal Cancer," in *Psychopharmacological Agents for the Terminally Ill and Bereaved*, I. K. Goldberg, S. Malitz, and A. H. Kutscher, eds. (New York and London: Columbia University Press, 1973).

14. Ibid., p. 102.
15. Her case is not unique. In a study of sixty cancer patients who were treated with LSD or a gentler psychedelic called DPT, 29 percent saw dramatic improvement, and 42 percent saw moderate improvement in their pain. Some cancer patients even returned to work after the experience. S. Grof et al., "LSD-Assisted Psychotherapy in Patients with Terminal Cancer," *International Pharmacopsychiatry* 8 (1973): 129–44.
16. Kurland et al., "Psychedelic Drug Assisted Psychotherapy," 113.

CHAPTER 7. SEARCHING FOR THE GOD SPOT

1. For example, a group of scientists in Sweden tried to replicate his findings, going so far as hiring the engineer who built Persinger's "God helmet" to make one for them. They then tested it on eighty-nine people, using a double-blind method. They concluded that some test subjects experienced a "sensed presence"—but it had nothing to do with the magnetic fields generated by the helmet. The relevant factor turned out to be personality. "Suggestible people"—those who were easily hypnotized, for example, or those who lived "a New Age lifestyle"—were far more likely to be transported by the helmet. See Pehr Granqvist et al., "Sensed Presence and Mystical Experiences Are Predicted by Suggestibility, Not by the Application of Transcranial Weak Complex Magnetic Fields," *Neuroscience Letters* 379 (2005): 1–6. The Swedes concluded that since people most responsive to the helmet score high on suggestibility, "placing the helmet on their heads in a sensory deprivation context might have the anticipated effects, whether or not the cord is plugged in" (p. 5).

   In response, Persinger reanalyzed data from 407 subjects and reiterated his claim that the magnetic configurations, not the subjects' exotic beliefs or suggestibility, were responsible for sensing a presence. But, he conceded, the subjects' histories of sensed presences experienced before the experiment were "moderately" correlated with exotic beliefs and temporal lobe sensitivity. See M. A. Persinger and S. A. Koren, "A Response to Granqvist et al.: 'Sensed presence and mystical experiences are predicted by suggestibility, not by the application of transcranial weak magnetic fields,'" *Neuroscience Letters* 380 (2005): 346–47.
2. For an excellent synopsis, see J. Saver and J. Rabin, "The Neural Substrates of Religious Experience," *Journal of Neuropsychiatry* 9 (1997): 498–510.

   Among the religious figures who supposedly had epilepsy are: Moses, Ezekiel (Jewish prophet), Saint Paul, Muhammad, Joan of Arc, Saint Catherine of Genoa, Saint Teresa of Ávila, Saint Catherine de' Ricci, Saint Thérèse de Lisieux, Emanuel Swedenborg (founder of the New Jerusalem Church), Ann Lee (leader of the Shaker movement), Joseph Smith (founder of the Church of Jesus Christ of Latter-day Saints, or Mormonism), Mary Baker Eddy (founder of Christian Science), Ellen G. White (founder of the Seventh-day Adventist Church), and Hieronymus Jaegen (German mystic).
3. Acts 9:3–5 (King James Version).
4. K. Dewhurst and A. W. Beard, "Sudden Religious Conversions in Temporal Lobe Epilepsy," *British Journal of Psychiatry* 117 (1970): 497–507.
5. W. G. Lennox, in *Epilepsy and Related Disorders,* vol. 2 (London: Churchill, 1960).
6. William James, *The Varieties of Religious Experience: A Study in Human Nature* (Cambridge, Mass.: Harvard University Press, 1985; originally published 1902), pp. 14–15.
7. This bothered John Hughes as well. Hughes has directed the Epilepsy Center at the University of Illinois Medical Center for thirty years. He is a Christian as well as a neurologist, a bit like being a zebra in a herd of horses: you *sort of* look like the others, but not really. Hughes researched forty-three alleged cases of epilepsy, from Pythagoras (born 582 B.C.) to Richard Burton (born A.D. 1925), searching for evidence of epilepsy.

   "How many of those people actually had epilepsy?" I asked him. "Zero," he said. "I looked up the literature and very carefully reviewed the history of those people and found that none of them had epilepsy." Some were diagnosed for silly reasons: one person had a "fit of spleen," which meant he was irritable, not epileptic. Martin Luther was taken into custody by the Catholic Church, and the phrase "Martin Luther's seizure" became his ticket to the annals of brain disorder. For others, Hughes said, the diagnosis did not match the symptoms. For example, Joan of Arc's visions stretched on for hours. Seizures last a couple of minutes.

Of course, Hughes's analysis has no more empirical heft than does that of neurologists who attribute religious fervor to complex partial seizures. His theory cannot be tested, either, since Moses and Paul are no longer available for a brain scan. But Hughes sees chicanery in the scientific community, which might have a hard time accommodating Saint Paul's experience within a materialist worldview. "I think it was some epileptologist who wanted to diminish Christianity by making Saint Paul's experience into a seizure," Hughes said. "Maybe it's the only thing scientists can do to try to put it in the context of the twenty-first century. But I'm very willing to see them as what I believe they were—truly deeply religious experiences."

8. W. Penfield and P. Perot, "The Brain's Record of Auditory and Visual Experience: A Final Discussion and Summary," *Brain* 86 (1963): 595–696.

9. Pierre Gloor and his colleagues found that surgical stimulation or spontaneous discharge in the hippocampus and amygdala—two areas deep in the temporal lobes—evoked memory fragments, dreamy states, and visual or auditory hallucinations. People reported that these events brought intense personal meaning, emotion, vibrations, fear, sudden insight, and mystical-like experiences. P. Gloor et al., "The Role of the Limbic System in Experimental Phenomena of Temporal Lobe Epilepsy," *Annals of Neurology* 12 (1982): 129–44.

10. W. Penfield, "The Role of the Temporal Cortex in Certain Psychical Phenomena," *Journal of Mental Science* 101 (1955): 458.

11. W. Penfield and T. Rasmussen, *The Cerebral Cortex of Man: A Clinical Study of Localization of Function* (New York: Macmillan, 1950), p. 174.

12. E. Slater and A. W. Beard, "The Schizophrenia-like Psychoses of Epilepsy: Psychiatric Aspects," *British Journal of Psychiatry* 109 (1963): 5–112; also "Discussion and Conclusions," ibid., 143–50.

13. K. Dewhurst and A. W. Beard, "Sudden Religious Conversions in Temporal Lobe Epilepsy," *British Journal of Psychiatry* 117 (1970): 497–507.

14. Not even people in psychiatric hospitals experience many spiritual seizures. Researchers in Japan studied 137 people with temporal lobe epilepsy and found that only three of them (2.2 percent) suffered seizures that were remotely religious in nature. Another study looked at 606 patients and found that only six had religious seizures. Akira Ogata and Taihei Miyakawa, "Religious Experiences in Epileptic Patients with a Focus on Ictus-Related Episodes," *Psychiatry and Clinical Neurosciences* 52 (1998): 321–25.

15. Norwegian researchers Bjørn Asheim Hansen and Eylert Brodtkorb studied eleven patients who experienced ecstatic seizures. Of those, five reported spiritual or religious experiential phenomena. Two felt contact with "an undescribable phenomenon" or a "divine power." One interpreted her ictal hallucinations to represent "the voice of God." Three subjects described the sensation of receiving deep messages during the seizures. Two felt that these experiences influenced their day-to-day lives between seizures; one interpreted the seizure experience as a prophecy with an objective of giving her life another dimension. B. A. Hansen and E. Brodtkorb, "Partial Epilepsy with 'Ecstatic' Seizures," *Epilepsy & Behavior* 4 (2003): 667–73.

16. The day of discovering the God spot, I suspect, will arrive far sooner than the Day of Judgment. Recently I watched Susan Bowyer, a medical physicist at Henry Ford Hospital in Detroit, create an image of a young woman's brain using MEG, or magnetoencephalography. MEG is brain scanning on steroids. Other types of brain-scanning technology, like fMRI

(functional magnetic resonance imaging), can record a static map of those areas of the brain that light up during a particular task. That would be like showing the route that O. J. Simpson took in his white Bronco when he fled the police. With the MEG, you can watch the brain work second by second, as if viewing the police chase live from a television news helicopter. Dr. Bowyer pointed to a computer that was recording the woman's brain as she performed a word task.

"I can tell you what your brain is doing from the minute you see the stimulus to the time you push a button and make a decision," she explained. "I can see it go from your visual cortex, to Wernicke's area for language, to the frontal for memory, going back to Broca's area before you say something, going on to the areas where you make the decision. So with MEG you can look at all the different areas and see which comes before which."

"Would the MEG, theoretically, be able to map a person's brain while she's having religious thoughts?" I asked.

"Probably," Dr. Bowyer said. "If you wanted to look at the religious areas [of the brain], you'd want to show a person maybe thirty religious icons, average them together, and see where they showed some activity."

"So would you be able to locate a God spot in the brain?"

Dr. Bowyer pondered the question. "You could show religious people the religious images and see where that evoked a response. And then take people who are atheists or nonreligious and show them the same images and see if they evoked any kind of emotional response. And maybe you could see if there is a God spot in the brain. Yeah," she said, nodding, "you could put together a study that might do that."

17. Saver and Rabin, "The Neural Substrates of Religious Experience," 499. See also D. Hay, "The Biology of God: What Is the Current Status of Hardy's Hypothesis?" *International Journal for the Psychology of Religion* 4 (1994): 1–23.

18. J. Wuerfel et al., "Religiosity Is Associated with Hippocampal but Not Amygdala Volumes in Patients with Refractory Epilepsy," *Journal of Neurology, Neurosurgery, and Psychiatry* 75 (2004): 640–42. See also L. Tebartz van Elst et al., "Psychopathological Profile in Patients with Severe Bilateral Hippocampal Atrophy and Temporal Lobe Epilepsy: Evidence in Support of the Geschwind Syndrome?" *Epilepsy & Behavior* 4 (2003): 291–97.

19. L. Tebartz van Elst et al., "Amygdala Abnormalities in Psychosis of Epilepsy: An MRI Study of Patients with Temporal Lobe Epilepsy," *Brain* 125 (2002): 593–624.

20. Ahahar Arzy et al., "Induction of an Illusory Shadow Person," *Nature* 443 (Sept. 21, 2006): 287.

21. M. A. Persinger and K. Makarec, "Complex Partial Epileptic Signs as a Continuum from Normals to Epileptics: Normative Data and Clinical Populations," *Journal of Clinical Psychology* 49, no. 1 (1993): 33–45.

CHAPTER 8. SPIRITUAL VIRTUOSOS

1. A. Newberg, E. D'Aquili, and V. Rause, *Why God Won't Go Away* (New York: Ballantine, 2001).

2. A. Newberg, and M. R. Waldman, *Why We Believe What We Believe* (New York: Free Press, 2006).

3. Since Newberg had already conducted brain scans of people meditating, engaging in centering prayer, chanting, and speaking in tongues, we thought we'd mix it up. Scott would pray for someone else in "intercessory prayer," and Newberg would take brain scans of both Scott and the recipient of his prayers. That guinea pig turned out to be me, as there weren't many people in Scott's church eager to have a radioactive tracer coursing through their veins. I thought this was pretty nifty, even though I am not fond of catheters. We were charting new territory—territory I will cover more fully in the next chapter. Would my brain "respond" to

Scott's prayers? To guard against anticipatory frontal lobe excitement in my brain—*He's praying for me now! Oh yeah, baby, I can feel it!*—we conducted two sessions. During one session, Scott would pray for me (prayer state). During the other, he would think about nothing in particular (baseline state). But I wouldn't know which was which. We would then see if my brain "responded" to his thoughts. This is called a "blind" study, where the subject does not know whether she's getting the placebo or the real thing—Scott's wandering mind or his prayers.

In a stroke of genius that would fell the whole endeavor, I had hired a sound man to record Scott's session, and in particular, the instructions Newberg gave him. (I was reporting the story for NPR as well, and thus needed the sound; but I could not be in the room when Scott prayed, or did not, as that would "unblind" the study.) All went well, until the intermission between sessions, when the engineer, clearly bored, began to chat with me.

"You know, it was interesting," he said helpfully. "After Scott had finished the session, he told me it was really hard not to—"

"Stop!" I said, covering my ears. *"Do not say another word!"*

"What's wrong?" he asked.

"Nothing. Don't worry about it."

The damage was done. I knew where that sentence was going: "—really hard not to pray." And thus the blind study was foiled.

I told Newberg that I knew that Scott had not prayed for me in the first session and he would pray for me during the next. Newberg stood there for a beat with a slightly frozen smile, then graciously reassured me that this sort of thing happens all the time in research. I was not comforted. I felt guilty and, frankly, peeved that it had taken a year for all the planets to align to allow Newberg, Scott, me, and the SPECT scanner to be free—only to be thwarted by a half-sentence. Newberg said he would continue the study for my sake: Scott would pray for someone else and get the scan, so we could see his brain in prayerful action. But the quirky side experiment that I was certain would change the world and defy materialism for good—that hope was dead.

As it turned out, Scott did pray for me, but my brain appeared to show no unusual activity.

Maybe another day.

4. S. Begley, "Your Brain on Religion: Mystic Visions or Brain Circuits at Work?" *Newsweek*, May 7, 2001.

5. Newberg did see one major difference in the two practices. The nuns were focusing on a word or phrase, and so the area of the brain that handles language lit up. The monks were focusing on a visual image, and so the visual areas of the brain lit up.

6. This story is recounted in Newberg and Waldman, *Why We Believe What We Believe*, pp. 198–99.

7. A. B. Newberg et al., "The Measurement of Regional Cerebral Blood Flow During Glossolalia: A Preliminary SPECT Study," *Psychiatry Research: Neuroimaging* 148 (2006): 67–71.

8. Romans 8:26 (New International Version).

9. See Nina P. Azari et al., "Neural Correlates of Religious Experience," *European Journal of Neuroscience* 13 (2001): 1649–52.

10. S. Begley, *Train Your Mind, Change Your Brain* (New York: Ballantine, 2007), p. 234.

11. R. J. Davidson et al., "Alterations in Brain and Immune Functions Brought by Mindfulness Meditation," *Psychosomatic Medicine* 65 (2003): 564–70.

12. In addition, both the newly minted meditators at Promega and the control group received a flu vaccine. As seen in other studies, the meditators developed more flu antibodies than did the control group, and suffered fewer flu symptoms. And the more meditation, the better the

immune system: those whose brain-wave activity tilted more leftward developed higher antibody titers.
13. F. Crick, *The Astonishing Hypothesis: The Scientific Search for the Soul* (London: Simon & Schuster, 1994), p. 3.

CHAPTER 9. OUT OF MY BODY OR OUT OF MY MIND?

1. Michael Sabom has published two books on near-death experiences: *Recollections of Death: A Medical Investigation* (New York: Harper & Row, 1982) and *Light and Death* (Grand Rapids, Mich.: Zondervan, 1998).
2. In one of the early studies, researchers claim that journals are filled with descriptions of out-of-body experiences that turn out to be accurate. In one study, a researcher analyzed 288 cases in which patients reported events that they could not have seen or heard with their physical senses. More compelling, in ninety-nine of those cases, the patients reported the event before it was verified. In other words, the experiencers could not have simply heard about it from someone else. See H. Hart, "ESP Projection: Spontaneous Cases and the Experimental Method," *Journal of the American Society for Psychical Research* 48 (1954): 121–46.
3. James said this in his Presidential Address to the (British) Society for Psychical Research on January 31, 1897; the address was published in *Proceedings of the Society for Psychical Research* 12 (1897): 5. My thanks to Bruce Greyson at the University of Virginia for locating this citation for me.
4. The surgery was reconstructed practically to the minute by Michael Sabom, who obtained Pam's records and wrote it up in *Light and Death*. I am grateful that he also spent considerable time with me in an interview, walking me through the process.
5. John 20:22. Then, fifty days later, when those same men were huddled together, seeking escape from the Romans, "suddenly a sound like the blowing of a violent wind came from heaven and filled the whole house where they were sitting. . . . All of them were filled with the Holy Spirit." Acts 2:2–4.
6. G. M. Woerlee, *Mortal Minds: The Biology of Near-Death Experiences* (New York: Prometheus, 2005).
7. This theory dates back to the 1950s, with the pioneering work of neurosurgeon Wilder Penfield. Before he conducted neurosurgery on an epileptic patient, Penfield routinely stimulated parts of the brain to figure out which parts to cut out and which to leave alone. His patients, who were awake (because the brain does not feel pain), could describe the sensations they were feeling, and where. In this way, he produced out-of-body–like phenomena. More recently, neurologist Orrin Devinsky and his colleagues investigated whether autoscopic experiences—body displacement similar to out-of-body experiences—occur when people suffer epileptic seizures. They studied ten of their own epileptic patients and thirty-three others who reported floating out of their bodies. Their conclusions: "Autoscopic seizures may be more common than is recognized; we found a 6.3% incidence in the patients we interviewed. The temporal lobe was involved in 18 (86%) of the 21 patients in whom the seizure focus could be identified." O. Devinsky et al., "Autoscopic Phenomena with Seizures," *Archives of Neurology* 46 (1989): 1080–88.
8. Or consider the case of a forty-three-year-old Swiss woman who came into the University Hospital in Geneva for a neurological evaluation. She had suffered for more than a decade from seizures that originated, it turned out, in the right temporal lobe. Neurologist Olaf Blanke opened her head, began stimulating parts of her brain, and suddenly, the woman felt herself leave her body. At first she reported that she was "sinking into the bed" or "falling from a great height." But when Blanke raised the voltage, he produced an out-of-body experience: "I see myself lying in bed from above," she said, "but I only see my legs and lower

trunk." Upping the voltage gave her the sensation of floating about six feet above the bed, but the next turn of the dial was less fun: "She reported that her legs appeared to be moving quickly towards her face, and took evasive action." I would, too, if my neurologist sicced my legs on me. Olaf Blanke et al., "Stimulating Illusory Own-Body Perceptions," *Nature* 419 (2002): 269–70. From this and from a later study involving five other people, Blanke theorized that a certain spot in the brain—where the temporal lobe and parietal lobe meet—was command central for out-of-body experiences. Scientists believe that this region of the brain orients the body in space, by integrating information about balance, touch, sight, and coordination. And if the messages going to that area are garbled—because of an epileptic seizure, or lesion, or artificial stimulation—then you could find yourself hovering near the ceiling, watching your own resuscitation. Olaf Blanke et al., "Out-of-Body Experience and Autoscopy of Neurological Origin," *Brain* 127 (2004): 243–58.

9. S. J. Blackmore, "A Psychological Theory of the Out-of-Body Experience," *Journal of Parapsychology* 48 (1984): 201–18.

10. K. Ring and S. Cooper, "Near-Death and Out-of-Body Experiences in the Blind: A Study of Apparent Eyeless Vision," *Journal of Near-Death Studies* 16 (1997): 101–47.

11. Mary Baker Eddy, *Science and Health, with Key to the Scriptures* (Boston: Christian Science Publishing Society, 1875), p. 486.

12. Aldous Huxley, *The Doors of Perception* (New York: Perennial Classics, 2004; originally published 1954), p. 24.

CHAPTER 10. ARE WE DEAD YET?

1. Raymond Moody, *Life After Life: An Investigation of a Phenomenon—Survival of Bodily Death* (New York: HarperCollins, 2001; originally published 1975).

2. In general, researchers say, the average near-death experience will pass through five fairly universal stages.

Stage one: Peace. Your heart stops, and while the emergency staff is thumping your chest and yelling orders, you are overwhelmed by peace, calm, contentment, no pain, no fear. Researchers figure between 60 and 85 percent of people who nearly die and return with some memory recall this profound peace. B. Greyson, "Incidence and Correlates of Near-Death Experiences in a Cardiac Care Unit," *General Hospital Psychiatry* 25 (2003): 269–76.

Stage two: Body separation, or out-of-body experience. You detach, float over your body, remember conversations verbatim; blind people have been reported to see the events that transpire. Sometimes you see outside the room, get a glimpse of the book your mother is reading in the waiting room, but generally you regard with detached puzzlement the train wreck that is your body. Estimates of how many people have out-of-body experiences varies widely, from 25 to 70 percent. Greyson, "Incidence and Correlates."

Stage three: Entering darkness. While a "tunnel" has become the popular symbol of the near-death experience, relatively few actually go through the tunnel. Between a quarter and a third say they enter darkness and then move toward a light. The first time, they break all speed limits, including the speed of light, but if they happen to be on their second near-death journey—as psychologist and researcher Scott Taylor told the audience at the Houston near-death conference—"they saunter. 'I've been here before, I know what's happening, I'm taking my time.'" When he said this, a number of people nodded their heads knowingly.

Stage four: Seeing the light. The light is always bright but never hurts the eyes. It comforts, and for the religious it is the physical manifestation of God. Between 16 and 70 percent of people see the light, depending on the study.

Stage five: Entering the light. This is a very busy time. Some people talk with a being, or beings, of light, ask them questions, and receive perfect answers. Others are greeted by dead friends and relatives (some of whom they have never met before). Some walk through meadows or a heavenly landscape. Others review their lives, generally with a "being" to guide them through the process. Many see a border—a fence, a door, a window, a bridge, a line in the sand, a river—and sense that if they cross over, they will never come back.

Finally they return: Sometimes they choose to return, drawn by unfinished business, such as children to raise, or a spouse to comfort. Others return kicking and screaming, like Pam Reynolds, who claimed she was pushed back into her body by her uncle. And always, the instant they return, so does the pain.

The percentages of each of these stages vary wildly, but meeting with the light and with dead relatives are the most common. Many near-death experiencers found being dead so lovely they were peeved when they came back.

"I was actually punched by a patient who said, 'You had no right to bring me back,'" Maggie Callanan, a hospice nurse and author of *Final Gifts* (New York: Bantam, 1992), told me in an interview. "And you know what I wanted to say to him? '*Do you know I missed lunch for you?*'"

3. One category of research examines whether these people are mad or not. In terms of psychological health, near-death experiencers are as sturdy as those who do not report these unusual death experiences. See Michael Sabom, *Recollections of Death: A Medical Investigation* (New York: Harper & Row, 1982); H. J. Irwin, *Flight of Mind: A Psychological Study of the Out-of-Body Experience* (Metuchen, N.J.: Scarecrow Press, 1985); and B. Greyson, "Near-Death Experiences Precipitated by Suicide Attempt: Lack of Influence of Psychopathology, Religion, and Expectations," *Journal of Near-Death Studies* 9 (1991): 183–88.

Near-death experiencers score the same in intelligence, mental health, and personality traits such as neuroticism (proneness to anxiety, fear, and depression) and extroversion (talkativeness, assertiveness, enthusiasm). See T. P. Locke and F. C. Shontz, "Personality Correlates of the Near-Death Experience: A Preliminary Study," *Journal of the American Society for Psychical Research* 77 (1983): 311–18. But look closely at this seemingly normal group and you find a few quirks. Bruce Greyson at the University of Virginia has collected stories from more than 1,000 people and found that those who report near-death experiences are more likely to report paranormal experiences, such as out-of-body experiences or vivid dreams. They tend to be putty in the hands of a hypnotist. One intriguing theory is that near-death experiencers enjoy vivid imaginations. For example, they score higher than average on being "fantasy prone"—that is, likely to report religious visions, ghosts, near-death experiences, and psychic abilities. See S. C. Wilson and T. X. Barber, "Vivid Fantasy and Hallucinatory Abilities in the Life Histories of Excellent Hypnotic Subjects ('Somnambules'): Preliminary Report with Female Subjects," in E. Klinger, ed., *Imagery*, vol. 2: *Concepts, Results, and Applications* (New York: Plenum, 1981), 133–49.

They also seem able to dissociate—replace an unpleasant reality with pleasurable fantasies to shield them from the emotional shock. Kenneth Ring and others have found that near-death experiencers report more abuse as a child than the average person. His theory is that these people learned early to cope with traumatic events by accessing "alternate realities"; later, faced with another traumatic event, these people are more likely to "flip" into that alternative consciousness and perceive things that others may not. See K. Ring, *The Omega Project: Near-Death Experiences, UFO Encounters, and Mind at Large* (New York: William Morrow, 1992).

They also tend toward "absorption." They often become focused on their thought life—particularly their imaginations—and can shut out the world. (Frankly, this sounds to me a whole lot like a reporter on deadline.) See A. Tellegen and G. Atkinson, "Openness to Absorbing and Self-Altering Experiences ('Absorption') a Trait Related to Hypnotic Susceptibility," *Journal of Near-Death Studies* 2 (1974): 132–39.

4. After a while, I came to think of the temporal lobe as the all-purpose lobe that skeptics

say is responsible for virtually every type of spiritual experience. Nonetheless, the temporal lobe, with its electrical spikes and auditory hallucinations and flashes of memory, plays a central role in explaining spiritual experience for many scientists. See M. L. Morse, D. Venecia, and J. Milstein, "Near-Death Experiences: A Neurophysiological Explanatory Model," *Journal of Near-Death Studies* 8 (1989): 45–53; also J. C. Saavedra-Aguilar and J. S. Gómez-Jeria, "A Neurobiological Model for Near-Death Experiences," *Journal of Near-Death Studies* 7 (1989): 205–22.

Willoughby Britton, a graduate student at the University of Arizona, compared twenty-three people who reported near-death experiences with twenty others who had not; he stuck electrodes on their heads overnight and recorded their brain activity as they slept. Five (21 percent) of the near-death experiencers recorded some mild anomalies in the left temporal lobe. Only one person (5 percent) in the control group exhibited these anomalies. They were not seizures, nor were they spikes; they wouldn't draw as much as a second glance from a neurologist looking for signs of epilepsy. But, Britton concluded, near-death experiences may involve the temporal lobe. See W. B. Britton and R. R. Bootzin, "Near-Death Experiences and the Temporal Lobe," *Psychological Science* 15, no. 4 (2004): 254–58.

5. G. M. Woerlee, "Cardiac Arrest and Near-Death Experiences," *Journal of Near-Death Studies* 22, no. 4 (2004): 245.

6. An intriguing scenario involved pilots in accelerator machines. They suffered blood loss to the head and felt some of the elements of a near-death experience: tunnel vision and bright lights, floating sensations, paralysis, "dreamlets" that featured family members. But these sensations were fragmented, and contained no life review or panoramic memory. See J. E. Whinnery, "Psychophysiologic Correlates of Unconsciousness and Near-Death Experiences," *Journal of Near-Death Studies* 15 (1997): 473–79.

Susan Blackmore advances a variation on the dying-brain theme. A British parapsychologist turned skeptic who has written several books on the phenomenon, Blackmore argues that oxygen deprivation in the visual areas of the brain creates lights and tunnels. The brain releases endorphins, creating feelings of peace and euphoria, similar to a runner's high. Seizures in the temporal lobe and limbic system trigger the life review, and out-of-body experiences arise when the brain attempts to reconstruct a plausible scenario of what happened, after the fact.

I tried to interview Blackmore on the subject, but was politely rebuffed. I caught a note of exasperation in her e-mail of Nov. 4, 2006: "I am sure you will understand when I say that I am not prepared to talk about NDEs any more. You may have seen on my website that I gave up all my research and media work on the paranormal and related topics many years ago. http://www.susanblackmore.co.uk/journalism/NS2000.html I am not going back on this. Media treatment of NDEs is (with very few exceptions) appalling. In every case I have been involved in the argument is polarised into 'Yes/No,' 'Life after death/boring science.' Over 30 years I failed completely to put over the alternative that NDEs are life-changing important experiences that tell us a lot about human nature and are therefore worth researching even though the idea of life after death is daft and entirely without evidence. Having been burned so many times," she concluded, "I am not going to talk about it any more."

Another controversial argument has been disavowed by its author, though it remains the favorite of many skeptics. It is this: Near-death experiences are the product of a hallucinating brain. Some scientists argue that the brain under stress creates a sort of organic LSD, triggering the visions, conversations, feelings, and story lines of near-death experiences. The strongest candidate is a neurotransmitter very similar to ketamine. Ketamine was used as an anesthetic on American soldiers during the Vietnam War, but was put on the shelf after they complained of bright lights and floating above their bodies. Dr. Karl Jansen suggested that the brain in distress might create a ketamine-like compound that would produce lights, and a sensation of floating. See K. L. R. Jansen, "The Ketamine Model of the Near-Death Experience: A

Central Role for the N-methyl-D-asparate Receptor," *Journal of Near-Death Studies* 16 (1997): 5–26. "It's pure speculation at this point," Bruce Greyson noted. "We don't know of any such compound. It's never been identified. So there may be a compound like this in the brain that *may* be produced under stress that *may* produce these effects. Well, yes, it *may* be. But if you look at ketamine experiences, they don't really mimic NDEs. There are certainly some things in common, but most ketamine experiences are terrifying, most people come back saying they do not want to go through that again, and they don't think of it as a real experience. Whereas virtually every near-death experiencer comes back saying, 'That was more real than me sitting here talking to you now, and I can't wait to go back.'"

Indeed, Jansen himself has shifted his position, suggesting that a ketamine-like chemical may be one trigger of authentic spiritual experience, while coming close to death may be another. He no longer rules out the possibility that people are catapulted—chemically or traumatically—into other worlds, and that these worlds might be, in fact, real. "I now believe that there most definitely is a soul that is independent of experience," he wrote. "It exists when we begin and may persist when we end. Ketamine is a door to a place we cannot normally get to; it is definitely not evidence that such a place does not exist." See K. L. R. Jansen, "Response to Commentaries on 'The Ketamine Model of the Near-Death Experience: A Central Role for the N-methyl-D-asparate Receptor,'" *Journal of Near-Death Studies* 16 (1997): 79–95. Blackmore insists that nothing, including consciousness, can separate from the body and survive. In fact, she asserted, "If . . . truly convincing paranormal events are documented then certainly the theory I have proposed will have to be overthrown." See Susan Blackmore, *Dying to Live: Near-Death Experiences* (New York: Prometheus, 1993), p. 262.

7. Peter Fenwick, "Science and Spirituality: A Challenge for the 21st Century," paper presented at International Association for Near-Death Studies (IANDS) 2004 Conference, Evanston, Illinois.

8. Mario Beauregard and Vincent Paquette, "Neural Correlates of a Mystical Experience in Carmelite Nuns," *Neuroscience Letters* 405 (2006): 186–90.

9. Specifically, one area that lit up in the brains of both nuns and near-death experiencers was a region of the temporal lobe called the middle temporal gyrus. Having considered reports by clinical neurologists and research on people with temporal lobe epilepsy, Beauregard said, "I hypothesize that this activation is related to the subjective perception of contacting a spiritual reality."

Another curiosity concerned the caudate nucleus, which is involved with feelings of intense love. "We just finished a study on unconditional love and we found out that this caudate nucleus was crucially involved with unconditional love. But it's also involved in other forms of love, like romantic love and maternal love. The major puzzle involved the parietal lobe, the area that helps you determine your body schema. In the nuns' brains, this area showed unusual activity, suggesting they were being absorbed into a larger being. The near-death experiencers showed no unusual activity at all, even though they talked about walking toward the light. "I'll have to think about this," was all Beauregard would say.

10. It is unlikely that a brain would rewire itself in an instant, Andrew Newberg told me. Newberg has puzzled over this, and says that if further research shows that a person's brain structure or brain-wave patterns do in fact behave differently as a result of a near-death or other experience, he can imagine a couple of mechanisms. Perhaps the jolting experience suddenly taps into unconscious areas of the brain; or perhaps it activates and strengthens weak connections in the brain, and the newly robust connections create a new pattern of brain activity that becomes the norm.

11. William James recognized this phenomenon a century ago. Mystical states, he wrote, in *The Varieties of Religious Experience* (Cambridge, Mass.: Harvard University Press, 1985; originally published 1902), allow the mystic to become one with the Absolute, and to be

aware of that oneness—a tradition that defied "clime or creed." "In Hinduism, in Neoplatonism, in Sufism, in Christian mysticism, in Whitmanism, we find the same recurring note, so that there is about mystical utterances an eternal unanimity which ought to make a critic stop and think, and which brings it about that the mystical classics have, as has been said, neither birthday nor native land" (p. 324).

CHAPTER 11. A NEW NAME FOR GOD

1. Albert Einstein, *Ideas and Opinions,* trans. Sonja Bargmann (New York: Dell, 1973), p. 255.
2. For an excellent account of Einstein's "God," see Walter Isaacson, *Einstein: His Life and Universe* (New York: Simon & Schuster, 2007), pp. 384–93.
3. Stephen Hawking, *A Brief History of Time* (New York: Bantam, 1988), p. 174.
4. Gregory Benford, "Leaping the Abyss: Stephen Hawking on Black Holes, Unified Field Theory and Marilyn Monroe," *Reason,* April 2002, p. 29.
5. Paul A. M. Dirac, "The Evolution of the Physicist's Picture of Nature," *Scientific American* 208 (May 1963): 53.
6. Max Planck, quoted in Charles C. Gillespie, ed., *Dictionary of Scientific Biography* (New York: Scribner, 1975), p. 15.
7. Anthony Flew, *There Is a God: How the World's Most Notorious Atheist Changed His Mind* (New York: Harper One, 2007), p. 155.
8. Freeman J. Dyson, *Disturbing the Universe* (New York: Harper & Row, 1979), p. 250.
9. Larry Dossey, *Recovering the Soul* (New York: Bantam, 1989).
10. The analogy breaks down slightly because, as Dossey and others have it, non-local mind possesses infinite information, not just boatloads of it, and non-local mind knows what is happening in the past, present, and future.
11. Dean Radin, *Entangled Minds* (New York: Paraview, 2006).
12. A. Aspect, P. Grangier, and G. Roger, "Experimental Realization of Einstein-Podolsky-Rosen-Bohm Gedankenexperiment: A New Violation of Bell's Inequalities," *Physical Review Letters* 49 (1992): 91–94. Many experiments have proved the same thing, separating particles by as much as thirty-one miles and still seeing entanglement.
13. Radin argues that each level is built on smaller ones: atoms are built from subatomic particles, molecules are built from atoms, chemicals are a lot of molecules, biology emerges from chemicals, society is a group of biological beings—all the way up to the level of the universe. He believes that science will find unexpected properties in biological systems (including ESP) that emerge from elementary forms of entanglement, just as water emerges from a unique combination of oxygen and hydrogen. From either individual element alone, you could not predict water.
14. Studies have been conducted also at Bastyr University in Washington, Washington University in St. Louis, the universities of Nevada and Hertfordshire, and University Hospital of Freiburg. See S. Schmidt, "Distant Intentionality and the Feeling of Being Stared At: Two Meta-analyses," *British Journal of Psychology* 95 (2004): 235–47.
15. Of the more than fifty studies, three of interest to me were:

L. J. Standish et al., "Electroencephalographic Evidence of Correlated Event-Related Signals Between the Brains of Spatially and Sensory Isolated Subjects," *Journal of Alternative and Complementary Medicine* 10 (2004): 307–14 (published by Mary Ann Liebert Publishers, Inc.). In five of the sixty subjects tested, the receiver's brain showed significantly higher brain activity when the sender was projecting an image. The chances that this would happen randomly to this number of people are more than 3,000 to 1. However, when researchers tried to replicate the results with the five successful subjects, only one showed a statistically significant "response."

D. Radin, "Event Related EEG Correlations Between Isolated Human Subjects," *Journal of Alternative and Complementary Medicine* 10 (2004): 315–23 (published by Mary Ann Liebert Publishers, Inc.). For three of the thirteen pairs of adult friends or relatives, the receiver's brain-wave activity jumped when the partner was sending positive intentions. On average, the receiver's EEG peaked 64 milliseconds after the sender's, then sloped downward, as did the sender's.

D. Radin and M. Schlitz, "Gut Feelings, Intuition, and Emotions: An Exploratory Study," *Journal of Alternative and Complementary Medicine* 11 (2005): 85–91 (published by Mary Ann Liebert Publishers, Inc.). In this case, involving twenty-six pairs of adults, one person sat in a shielded room, while another tried to evoke positive, negative, calming, or neutral responses. There seemed to be a "response" in the EEG of the receiver when the sender sent positive or negative emotions. The odds against these being chance findings were 167 to 1 and 1,100 to 1, respectively.

16. fMRI technology is far more expensive (about $1,000 per brain scan), and few researchers have access to these machines. Therefore, few studies have gone this route. The results, while mixed, have been suggestive that this is an area ripe for research. See L. J. Standish, "Evidence of Correlated Functional MRI Signals Between Distant Human Subjects," *Alternative Therapies* 9 (2003): 122–28. In one pair that was tested—a man and woman who had been colleagues for two years—when the man was sending images to the woman lying in the brain scanner, her brain lit up, or activated in areas 18 and 19 of the visual cortex. This is the region of the brain that is activated when someone directly sees an object.

17. J. Achterberg, "Evidence for Correlation Between Distant Intentionality and Brain Function in Recipients: A Functional Magnetic Imaging Analysis," *Journal of Alternative and Complementary Medicine* 11 (2005): 965–71 (published by Mary Ann Liebert Publishers, Inc.).

18. D. Radin, "Compassionate Intention as a Therapeutic Intervention by Partners of Cancer Patients: Effects of Distant Intention on the Automatic Nervous System," *Explore* 4 (2008): 235–43.

19. I asked Schlitz if she had found pairs other than bonded couples who excelled at these tests. She nodded.

"We see that they are typically people who come from three sets of trainings," she said. "They are meditators. They're martial artists. Or they're classically trained musicians. So you might ask the question, what do those three populations have in common? Well, they have in common both intention and attention training." A meditator trains his brain to be still and highly focused. An Aikido master learns to focus on the opponent in front of him and movement on the periphery, with "eyes on the back of his head." The same is true for a classically trained musician, she observed. "Their brains are actually different from a person who hasn't been trained that way. And one of the things a musician can do, for example, is attend to their own line in the symphony and stay very focused on a particular melody that they're doing, and at the same time they have the larger capacity to track the whole symphony as it's performing. So there's something about that focused attention combined with this more generalized intention." Having seen how meditation literally molds the brain, I was hardly surprised by this finding. It seemed to add another straw to the mounting pile of evidence that the trained brain has a capacity to glean information and dimensions that the flabby or distracted brain cannot.

20. Specifically, when the "senders" (such as J.D.) saw the image of their loved ones on the screen and began to think about them, certain things happened: for five seconds, their brain waves spiked, as did their heart rate and sweat-gland activity, and their blood flowed away from their fingertips, which happens when people gear up to do a task like focusing their attention. Then, halfway through, the process reversed as they began to relax. That much was

predictable. But what gave the researchers pause was the response of the "receivers" (such as Teena) in the soundproof, electromagnetically sealed room. The receivers mimicked their partners' physiology within a few milliseconds, becoming aroused and then relaxing toward the end of the ten seconds. One curious outcome, Radin said, involved breathing. "At the end of the sending period, the sender typically does a big exhalation, because they've been holding their breath for the ten seconds. There's also a big exhalation for the receiver at the same time, even though they're not holding their breath." He laughed. "I didn't expect that."

CHAPTER 12. PARADIGM SHIFTS

1. My colleagues hailed from *The New York Times*, *The Washington Post*, *The Philadelphia Inquirer*, *USA Today*, *Newsday*, ABC News, the BBC, and *New Scientist* magazine. Another journalist made his living authoring popular science books. This was the sharpest and most magical group of people I have known, their intellect rivaled only by their capacity to laugh and to drink.

2. Richard Dawkins, *The God Delusion* (New York: Bantam, 2006).

3. Thomas Kuhn, *The Structure of Scientific Revolutions* (Chicago: University of Chicago Press, 1996, 3rd edition; originally published 1962).

4. This quotation has repeatedly, and wrongly, been attributed to Kuhn. In fact, the words are those of a science writer reviewing Kuhn's book: Nicholas Wade, "Thomas S. Kuhn: Revolutionary Theorist of Science," *Science* 197, no. 4299 (1997): 144.

5. Kuhn, *Structure*, p. 6.

6. Ibid., p. 5.

7. Darren Staloff, "James's Pragmatism," a lecture in the series "Great Minds of the Western Intellectual Tradition," produced by the Teaching Company.

8. William James, *Pragmatism* (New York: Longmans, Green, 1916), p. 107 (italics mine).

9. Mary Baker Eddy, *Science and Health, with Key to the Scriptures* (Boston: Christian Science Publishing Society, 1875), p. 587.

Index